SpringerBriefs in Computer Science

For further volumes:
http://www.springer.com/series/10028

Keijo Haataja · Konstantin Hyppönen
Sanna Pasanen · Pekka Toivanen

Bluetooth Security Attacks

Comparative Analysis, Attacks, and Countermeasures

 Springer

Keijo Haataja
Konstantin Hyppönen
Sanna Pasanen
Pekka Toivanen
School of Computing
University of Eastern Finland
Kuopio
Finland

ISSN 2191-5768   ISSN 2191-5776   (electronic)
ISBN 978-3-642-40645-4   ISBN 978-3-642-40646-1   (eBook)
DOI 10.1007/978-3-642-40646-1
Springer Heidelberg New York Dordrecht London

Library of Congress Control Number: 2013949249

Printed on acid-free paper

Springer is part of Springer Science+Business Media (www.springer.com)

Preface

We gratefully acknowledge Elena Trichina, Martti Penttonen, and Tapio Grönfors for their guidance and supervision during the work presented in Haataja's Ph.D. thesis [1]. We also thank Niina Päivinen for her cooperation on RF-Fingerprint related research work [2]. This research work was funded by the European Union Artemis project Design, Monitoring, and Operation of Adaptive Networked Embedded Systems (DEMANES).

References

1. K. Haataja, Security Threats and Countermeasures in Bluetooth-Enabled Systems, Ph.D. Diss., University of Kuopio, Department of Computer Science, 2009
2. S. Pasanen, K. Haataja, N. Päivinen, P. Toivanen, New Efficient RF Fingerprint-Based Security Solution for Bluetooth Secure Simple Pairing, in *Proceedings of the 43rd IEEE Hawaii International Conference on System Sciences*, Koloa, Kauai, 5–8 Jan 2010

Contents

Chapter 1
Introduction

The use of wireless communication systems and their interconnections via networks have grown rapidly in recent years. Because radio frequency (RF) waves can penetrate obstacles, wireless devices can communicate with no direct line of sight between them. This makes RF communication easier to use than wired or infrared communication, but it also makes eavesdropping easier. Moreover, it is easier to disrupt and jam wireless RF communication than wired communication. Because wireless RF communication can suffer from these new threats, additional countermeasures are needed to protect against them.

Bluetooth [1] is a technology for short-range wireless data and real-time two-way audio/video transfer providing data rates up to 24 Mb/s. *Connection types* define the ways Bluetooth devices can exchange data. Bluetooth has three connection types: *Asynchronous Connection-Less (ACL), Synchronous Connection-Oriented (SCO)*, and *Extended SCO (eSCO)*.

ACL links are used for symmetric or asymmetric data transfer. Retransmission of packets is used to ensure the integrity of data. *SCO links* are symmetric and are used for transferring real-time two-way voice. Retransmission of voice packets is not used. Therefore, when the channel Bit Error Rate (BER) is high, voice can be distorted. *eSCO links* are also symmetric and are used for transferring real-time two-way voice. Retransmission of packets is used to ensure the integrity of data (voice). Because retransmission of packets is used, eSCO links can also carry data packets. However, they are mainly used for transferring real-time two-way voice. Bluetooth 1.2 (or later) devices can use eSCO links, but they must also support SCO links to provide backward-compatibility.

Bluetooth operates at 2.4 GHz frequency in the free Industrial, Scientific, and Medical (ISM) band. Bluetooth devices that communicate with each other form a *piconet*. The device that initiates a connection is the piconet *master* and all other devices within that piconet are *slaves*. All communication within a piconet goes through the piconet master. The clock of the piconet master and frequency-hopping information are used to synchronize the piconet slaves with the master. Two or more piconets together form a *scatternet*, which can be used to eliminate Bluetooth range

K. Haataja et al., *Bluetooth Security Attacks*, SpringerBriefs in Computer Science, DOI: 10.1007/978-3-642-40646-1_1, © The Author(s) 2013

restrictions. A scatternet environment requires that different piconets must have a common device, called a *scatternet member*, to relay data between the piconets.

Many kinds of Bluetooth devices, such as laptops, PCs, mice, keyboards, printers, mobile phones, headsets, and hands-free devices, are widely used all over the world. Moreover, in many countries, a hands-free device or headset connected to a mobile phone is the only form of voice communication device permitted in moving vehicles for safety reasons. Therefore, the markets for easy-to-use wireless Bluetooth headsets and hands-free devices are huge!

Already in 2006, the one billionth Bluetooth device was shipped [2]. Less than 5 years later in 2011, the four billionth Bluetooth device was shipped [3], and the volume is expected to increase rapidly in the near future. According to In-Stat, the eight billionth Bluetooth device is expected to be shipped by the end of 2013 [4]. Therefore, it is very important to keep Bluetooth security issues up to date.

Our results: In this book, we explain the reasons for Bluetooth network vulnerabilities and provide a literature-review-based comparative analysis of Bluetooth security attacks over the past 10 years (2001–2011), including our own Bluetooth security attacks. In addition, we explain countermeasures against these attacks based on a literature review and propose a new practical countermeasure for Bluetooth Secure Simple Pairing (SSP). We propose a new practical attack that works against all existing Bluetooth versions. Furthermore, we present some new ideas that will be used in our future research work.

The rest of the book is organized as follows. Chapter 2 provides an overview of Bluetooth security. Reasons for Bluetooth network vulnerabilities are explained in Chap. 3. Chapter 4 provides a literature-review-based comparative analysis of Bluetooth security attacks over the past 10 years (2001–2011): the attacks are designed against Bluetooth versions up to 2.0 + EDR (Enhanced Data Rate), but some of them (especially Denial-of-Service attacks) also work against all existing Bluetooth versions, i.e., Bluetooth versions 1.0A–4.0. Since Man-In-The-Middle (MITM) attacks are also possible and dangerous against the latest SSP-enabled Bluetooth versions (i.e., Bluetooth versions 2.1 + EDR − 4.0), MITM attacks on Bluetooth are explained in Chap. 5, which also provides a comparative analysis of all existing Bluetooth MITM attacks: MITM attacks, against "old" Bluetooth versions up to 2.0 + EDR as well as against "new" SSP-enabled Bluetooth versions, over the past 10 years (2001– 2011) are explained and analyzed. Chapter 6 explains the existing countermeasures against these attacks based on a literature review and proposes a new practical countermeasure for Bluetooth SSP. A new practical attack that works against all existing Bluetooth versions is proposed in Chap. 7. Finally, Chap. 8 concludes the book and sketches future work.

Chapter 2
Overview of Bluetooth Security

The basic Bluetooth security configuration is done by the user who decides how a Bluetooth device will implement its connectability and discoverability options. The different combinations of connectability and discoverability capabilities can be divided into three categories, or *security levels* [1, 2].

1. *Silent:* The device will never accept any connections. It simply monitors Bluetooth traffic.
2. *Private:* The device cannot be discovered, i.e., it is a so-called *non-discoverable device*. Connections will be accepted only if the *Bluetooth Device Address (BD_ADDR)* is known to the prospective master. A 48-bit BD_ADDR is normally unique and refers globally to only one individual Bluetooth device.
3. *Public:* The device can be both discovered and connected to. It is therefore called a *discoverable device*.

The 48-bit BD_ADDR is divided into three parts: the 16-bit Nonsignificant Address Part (NAP), the 8-bit Upper Address Part (UAP), and the 24-bit Lower Address Part (LAP). The first three bytes of BD_ADDR (NAP and UAP) refer to the manufacturer of the Bluetooth chip and represent *company_id*. The last three bytes of BD_ADDR (LAP), called *company_assigned*, are used more or less randomly in different Bluetooth device models. Company_id values are public information and are listed in the Institute of Electrical and Electronics Engineers' (IEEE's) Organizationally Unique Identifier (OUI) database [2, 5].

There are also four different *security modes* that a device can implement. In Bluetooth technology, a device can be in only one of the following security modes at a time [1, 2]:

1. *Nonsecure:* The Bluetooth device does not initiate any security measures.
2. *Service-level enforced security mode:* Two Bluetooth devices can establish a nonsecure ACL link. Security procedures, namely authentication, authorization, and optional encryption, are initiated when a Logical Link Control and Adaptation Protocol (L2CAP) Connection-Oriented (CO) or an L2CAP Connection-Less (CL) channel request is made.

K. Haataja et al., *Bluetooth Security Attacks*, SpringerBriefs in Computer Science, DOI: 10.1007/978-3-642-40646-1_2, © The Author(s) 2013

3. *Link-level enforced security mode:* Security procedures are initiated when an ACL link is established.
4. *Service-level enforced security mode:* This mode is similar to mode 2, except that only Bluetooth devices using SSP can use it, i.e., only Bluetooth 2.1+EDR or later devices can use this security mode.

Authentication is used for proving the identity of one piconet device to another. The results of authentication are used for determining the client's *authorization level*, which can be implemented in many different ways: for example, access can be granted to all services, only to a subset of services, or to some services while other services require additional authentication. *Encryption* is used for encoding the information being exchanged between Bluetooth devices in such a way that eavesdroppers cannot read its contents [2].

Bluetooth uses *Secure And Fast Encryption Routine + (SAFER+)* [6] with a 128-bit key as an algorithm for authentication and key generation in Bluetooth versions up to 3.0+HS (High Speed), while Bluetooth 4.0 (i.e. Bluetooth Low Energy) replaces SAFER+ with the more secure 128-bit *Advanced Encryption Standard (AES)* [1, 2, 7].

SAFER+ [6] was developed by Massey et al. in 1998. It was submitted as a candidate for the AES contest, but the cipher was not selected as a finalist. SAFER+ is a block cipher with the following main features. It has a block size of 128 bits and three different key lengths (128, 192, and 256 bits). SAFER+ consists of nine phases (eight identical rounds and the output transformation) and a Key Scheduling Algorithm (KSA) in the following way. KSA produces 17 different 128-bit subkeys. Each round uses two subkeys and a 128-bit input word from the previous round to calculate a 128-bit word that is a new input word for the next round. The last subkey is used in the output transformation, which is a simple bitwise XOR of the last round's output with the last subkey. Although some optimizations for faster breaking of SAFER+ exist (for example, in [8, 9]), it is still considered quite secure [1, 2, 6, 8, 9].

AES [7] was published by the National Institute of Standards and Technology (NIST) in 2001 after the evaluation process of the AES contest. Rijndael was the winner of the contest and NIST selected it as the algorithm for AES. AES is a symmetric block cipher that is intended to replace the Data Encryption Standard (DES) as the approved standard for a wide range of applications, but this process will take many years. NIST anticipates that the Triple Data Encryption Standard (3DES) will remain an approved algorithm for the foreseeable future, at least for U.S. government use. AES encryption consists of 10–14 rounds in which data blocks are processed step-by-step in the following way (except the final round; it is noteworthy that AES decryption is symmetric to AES encryption) [1, 2, 7, 10, 11]:

1. *Byte substitution:* Byte substitution uses an S-box to perform a byte-by-byte substitution of the block.
2. *Row shifting:* Row shifting is a simple permutation.

3. *Column mixing:* Column mixing is a substitution which makes use of arithmetic over $GF(2^8)$. Galois Field $GF(2^8)$ is a finite field of 256 elements, which can be denoted by strings of eight bits or by hexadecimal notation.
4. *Round key adding:* Round key adding is a simple bitwise XOR of the current block with a portion of the expanded key.

The final round of AES encryption (and AES decryption) is slightly different [1, 2, 7, 10, 11]:

1. *Byte substitution.*
2. *Row shifting.*
3. *Round key adding.*

AES is considered secure, it is very fast and compact (about 1 kB of code), its block size is a multiple of 32 (typically 128 bits), its key length is also multiples of 32 (typically 128, 192, or 256 bits), and it has a very neat algebraic description [2, 7, 10, 11].

Because Bluetooth is a wireless communication system, there is always a possibility that the transmission could be deliberately jammed or intercepted, or that false or modified information could be passed to the piconet devices [2].

Bluetooth security is based on building a chain of events, none of which should provide meaningful information to an eavesdropper. All events must occur in a specific sequence for security to be set up successfully [1, 2].

In order for two Bluetooth devices to start communicating, a procedure called *pairing* must be performed. As a result of pairing, two devices form a trusted pair and establish a link key which is used later on for creating a data encryption key for each session [2].

In Bluetooth versions up to 2.0+EDR, pairing is based exclusively on the fact that both devices share the same *Personal Identification Number (PIN)*, or *passkey*, that is used for generating several 128-bit keys as Fig. 2.1 illustrates. When the user enters the same passkey in both devices, the devices generate the same shared secret which is used for authentication and encryption of traffic exchanged by them [1, 2].

An *initialization key* (K_{init}) is generated when Bluetooth devices meet for the first time and it is used for securing the generation of other more secure 128-bit keys, which are generated during the next phases of the security chain of events. K_{init} is derived from a 128-bit pseudorandom number IN_RAND, an L-byte ($1 \leq L \leq 16$) PIN code, and a BD_ADDR. It is worth noting that IN_RAND is sent via air in unencrypted form [1, 2].

The output of a certain key generation function can be expressed in terms of the function itself and its inputs. K_{init} is produced in both devices using the formula $K_{init} = E_{22}(PIN',L',IN\_RAND)$. The PIN code and its length L are modified into two different quantities called PIN' and L' before they are sent to the E_{22} function. If the PIN is smaller than 16 bytes, it is augmented by appending bytes from the device's BD_ADDR until either PIN' reaches a total length of 16 bytes or the entire BD_ADDR is appended, whichever comes first. If one device has a fixed PIN code,

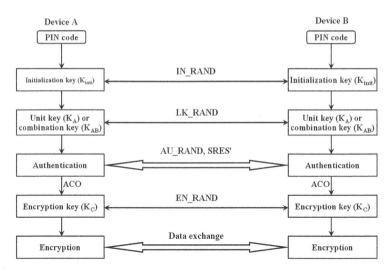

Fig. 2.1 Summary of Bluetooth security operations [1, 2]

the BD_ADDR of the other device is used. If both devices can support a variable PIN code, the BD_ADDR of the device that received IN_RAND is used [1, 2].

K_{init} is used to encrypt a 128-bit pseudorandom number (RAND or LK_ RAND), i.e., RAND $\oplus K_{init}$ or LK_RAND $\oplus K_{init}$, exchanged in the next phase of the security chain of events when a link key (a unit key or a combination key) is generated [1, 2].

A *unit key* (K_A) is produced from the information of only one device (device A) using the formula $K_A = E_{21}(BD\_ADDR_A, RAND_A)$. Device A encrypts K_A with K_{init}, i.e., $K_A \oplus K_{init}$, and sends it to device B. Device B decrypts K_A with K_{init}, i.e., $(K_A \oplus K_{init}) \oplus K_{init} = K_A$, and now both devices have the same K_A as a link key [1, 2].

A Bluetooth device that uses a unit key has only one key to use for all of its connections. This means that the same key is shared with all other trusted Bluetooth devices. In addition, any trusted Bluetooth device using the same unit key can eavesdrop on any traffic between two other Bluetooth devices that share the same unit key. Moreover, any trusted Bluetooth device using the same unit key can impersonate the target device just by duplicating its BD_ADDR. Thus, only devices that have limited resources, i.e., no memory to store several keys, should use K_A, because it provides only a low level of security. Therefore, Bluetooth specifications do not recommend using K_A anymore. More information about unit key weaknesses can be found in [1, 2, 12].

A *combination key* (K_{AB}) is dependent on two devices and therefore it is derived from the information of both devices. K_{AB} is produced in both devices using the formula $K_{AB} = E_{21}(BD\_ADDR_A, LK\_RAND_A) \oplus E_{21}(BD\_ADDR_B, LK\_RAND_B)$. It is worth noting that generating K_{AB} is nothing more than a simple bitwise XOR between two unit keys, i.e. $K_{AB} = K_A \oplus K_B$. Each device can produce its own unit

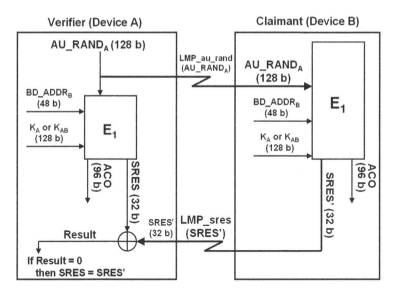

Fig. 2.2 Bluetooth challenge-response authentication [1, 2]

key and each device also has the BD_ADDR of the other device. Therefore, two devices have to exchange only their respective pseudorandom numbers in order to produce each other's unit key [1, 2].

Device A encrypts LK_RAND$_A$ with the current key K, i.e., LK_RAND$_A \oplus$ K where K can be the K$_{init}$, the K$_A$, or the K$_{AB}$ that was created earlier, and sends it to device B. K is K$_{init}$ if the devices create a link key for the first time together. K is K$_A$ if the link key is being upgraded to a combination key, and it is K$_{AB}$ if the link key is being changed. Device B decrypts LK_RAND$_A$ with K, i.e., (LK_RAND$_A \oplus$ K) $\oplus$ K = LK_RAND$_A$, and can now produce K$_A$. Correspondingly, device B encrypts LK_RAND$_B$ with K, i.e., LK_RAND$_B \oplus$ K, and sends it to device A. Device A decrypts LK_RAND$_B$ with K, i.e., (LK_RAND$_B \oplus$ K) $\oplus$ K = LK_RAND$_B$, and produces K$_B$. Finally, both devices can produce K$_{AB}$ by XORing K$_A$ with K$_B$, i.e., K$_{AB}$ = K$_A \oplus$ K$_B$ [1, 2].

The next phase of the security chain of events is the *challenge-response authentication* in which a claimant's knowledge of a secret link key is checked, as Fig. 2.2 illustrates. During each authentication, a new 128-bit pseudorandom number AU_RAND is exchanged via air in unencrypted form. Other inputs to the authentication function E$_1$ are the BD_ADDR of the claimant and the current link key (K$_A$ or K$_{AB}$) [1, 2].

A 32-bit result *(SRES, Signed Response)* and a 96-bit result *(ACO, Authenticated Ciphering Offset)* are produced in both devices by the E$_1$(AU_RAND$_A$, BD_ADDR$_B$,Link key) function, where the Link key is K$_A$ or K$_{AB}$. The claimant sends SRES', i.e., the SRES value produced by the claimant, via air in unencrypted form to the verifier. The verifier compares the generated SRES value with the received

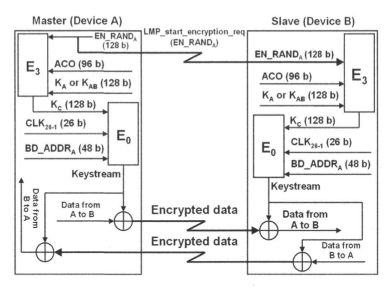

Fig. 2.3 Bluetooth data encryption [1, 2]

SRES value, and if these values match, the authentication is completed successfully. The ACO is used in the next phase of the security chain of events when an encryption key is generated [1, 2].

It is worth noting that SRES and SRES' are 32-bit values, not 128-bit values. The 32-bit SRES provides reasonable protection against an attacker who is trying to guess the value, while it also reduces the chance that the PIN code will be compromised by an attacker who has somehow determined the correct SRES value [2].

Figure 2.3 illustrates Bluetooth data encryption between two Bluetooth devices. The ACO, the current link key (K_A or K_{AB}) and a 128-bit pseudorandom number EN_RAND are the inputs to the encryption key generation function E_3 that is used for generating an *encryption key* (K_C). The master (device A) generates EN_RAND and sends it to the slave (device B) via air in unencrypted form. K_C is produced in both devices using the formula $K_C = E_3(EN\_RAND_A, ACO, Link key)$, where the Link key is K_A or K_{AB} [1, 2].

The keystream generator function E_0 (see Fig. 2.3) makes symmetric encryption possible by generating the same cipher bit stream, or *keystream* (also referred to as a *running key*), in both devices. The inputs to the E_0 function are K_C, the BD_ADDR of the master (BD_ADDR$_A$), and the 26 bits of the master's real-time clock (CLK$_{26-1}$). The keystream is generated by the $E_0(K_C, CLK_{26-1}, BD\_ADDR_A)$ function that is reinitialized for every new sent or received Baseband packet, i.e., CLK$_{26-1}$ is updated for every new Baseband packet. This means that inputs to the E_0 function are never identical longer than the lifetime of one Baseband packet and therefore a new keystream is generated for every new Baseband packet [1, 2].

The sender encrypts the plaintext by XORing it with the keystream, i.e., Plaintext $\oplus$ Keystream = Ciphertext, and sends the produced ciphertext to the receiver. The receiver decrypts the ciphertext by XORing it with the same keystream, i.e., Ciphertext $\oplus$ Keystream = (Plaintext $\oplus$ Keystream) $\oplus$ Keystream = Plaintext. It is worth noting that only the payload of the Bluetooth Baseband packet is encrypted (not an access code or a header), and therefore an attacker cannot use the regularly repeating information (that is easy for the attacker to guess) of the access code and the header in order to facilitate a cryptanalysis of the cipher [2].

As already discussed in this chapter, the PIN is the only source of entropy for the shared secret in Bluetooth versions up to 2.0+EDR. As the PINs often contain only four decimal digits, the strength of the resulting keys is not enough for protection against passive eavesdropping on communication. Even with longer 16-character alphanumeric PINs, full protection against active eavesdropping cannot be achieved: it has been shown that MITM attacks on Bluetooth communications (versions up to 2.0+EDR) can be performed [2, 13–15].

Let us assume that Alice and Bob are communicating with each other and they want to secure their communication by using some public-key encryption method. In a *MITM attack*, Mallory (an attacker) intrudes between Alice and Bob. Mallory can eavesdrop on messages, modify messages, delete messages, and generate new messages between Alice and Bob in such a way that his presence is unrevealed, i.e., Alice and Bob do not know that the link between them is compromised by Mallory. Mallory is also able to imitate Bob when talking to Alice and vice versa. This simple example of a MITM attack works in the following way [2, 10, 16]:

1. Alice sends her public key to Bob, but Mallory is able to intercept it. Mallory sends Bob his own public key for which he has the matching private key. Now Bob wrongly thinks that he has Alice's public key.
2. Bob sends his public key to Alice, but Mallory is able to intercept it. Mallory sends Alice his own public key for which he has the matching private key. Now Alice wrongly thinks that she has Bob's public key.
3. Alice sends Bob a message encrypted with Mallory's public key, but Mallory is able to intercept it. Mallory decrypts the message with his private key, keeps a copy of the message, re-encrypts the message with Bob's public key, and sends the message to Bob. Now Bob wrongly thinks that the message came directly from Alice.
4. Bob sends Alice a message encrypted with Mallory's public key, but Mallory is able to intercept it. Mallory decrypts the message with his private key, keeps a copy of the message, re-encrypts the message with Alice's public key, and sends the message to Alice. Now Alice wrongly thinks that the message came directly from Bob.

Even if the public keys of Alice and Bob are stored on a database, a MITM attack will work. Mallory can intercept Alice's (or Bob's) database inquiry and substitute his own public key for Bob's (or Alice's) public key. He can also somehow break into the database and substitute his key for both Alice's public key and Bob's public key. A MITM attack works, because Alice and Bob have no way to verify that they are truly using each other's correct public keys. If Mallory does not cause any noticeable delays to the communication, Alice and Bob have no idea that Mallory has intruded between them [2, 10, 11, 16–18].

Without any verification of the public keys, MITM attacks are generally possible (in principle) against any message sent using public-key technology. One solution to this problem is to use *public key certificates* (also referred to as *digital identity certificates*) [17], which use digital signatures to bind together public keys with the information of their respective users, i.e., information such as the name of the user, the address of the user, and so on. Each user is associated with a trusted authority, a *Certification Authority (CA)*, and each certificate is created by such a CA. A certificate establishes a verifiable connection between the user and his public keys. The users know their CA's public key and therefore they can verify the signatures of their CA. The certificate is stored in a directory. Only the CA is allowed to write in this directory, but all users of the CA can read the information in the directory [2, 10, 11, 16–18].

Defences against MITM attacks use authentication techniques which are based on *public key certificates*, *two-way authentication* (also referred to as *mutual authentication*), *secret keys*, *passwords*, and other methods (such as *voice recognition* and *other biometrics*) [2, 10, 11, 16–18].

Bluetooth versions 2.1+EDR, 3.0+HS, and 4.0 add a new specification for the pairing procedure, namely SSP [1]. Its main goal is to improve the security of pairing by providing protection against passive eavesdropping and MITM attacks [1, 2].

Instead of using (often short) passkeys as the only source of entropy for building the link keys, SSP employs Elliptic Curve Diffie-Hellman (ECDH) public-key cryptography. To construct the link key, devices use public-private key pairs, a number of nonces, and Bluetooth addresses of the devices. Passive eavesdropping is effectively thwarted by SSP, as running an exhaustive search on a private key with approximately 95 bits of entropy is currently considered to be infeasible in reasonable time [1, 2].

In order to provide protection against MITM attacks, SSP either uses an OOB channel (e.g., Near Field Communication, NFC), or asks for the user's help: for example, when both devices have displays and keyboards, the user is asked to compare two six-digit numbers. Such a comparison can also be thought of as an OOB channel which is not controlled by the MITM. If the values used in the pairing process have been tampered with by the MITM, the six-digit integrity checksums will differ with probability 0.999999 [1, 2].

SSP uses four *association models*. In addition to the two association models mentioned previously, *OOB* and *Numeric Comparison*, models named *Passkey Entry* and *Just Works* are defined [1, 2].

The Passkey Entry association model is used in cases when one device has input capability, but no screen that can display six digits. A six-digit checksum is shown to the user on the device that has output capability, and the user is asked to enter it on the device with input capability. The Passkey Entry association model is also used if both devices have input but not output capabilities. In this case the user chooses a 6-digit checksum and enters it in both devices. Finally, if at least one of the devices has neither input nor output capability, and an OOB cannot be used, the

Table 2.1 Device capabilities and SSP association models [1, 2]

Device 1	Device 2	Association model
DisplayYesNo	DisplayYesNo	Numeric comparison[a]
	DisplayOnly	Numeric comparison
	KeyboardOnly	Passkey Entry[a]
	NoInputNoOutput	Just works
DisplayOnly	DisplayOnly	Numeric comparison
	KeyboardOnly	Passkey entry[a]
	NoInputNoOutput	Just Works
KeyboardOnly	KeyboardOnly	Passkey entry[a]
	NoInputNoOutput	Just works
NoInputNoOutput	NoInputNoOutput	Just works

[a] The resulting link key is considered *authenticated*

Just Works association model is used. In this model the user is not asked to perform any operations on numbers: instead, the device may simply ask the user to accept the connection [1, 2].

The choice of the association model depending on the device capabilities is shown in Table 2.1. *DisplayYesNo* indicates that the device has a display and at least two buttons that are mapped to "yes" and "no": using the buttons the user can either accept the connection or decline it. Other notation in the table is self-explanatory [1, 2].

SSP is comprised of six phases: [1, 2]

1. *Capabilities exchange:* The devices that have never met before or want to perform re-pairing for some reason, first exchange their Input/Output (IO) capabilities (see Table 2.1) to determine the proper association model to be used.
2. *Public key exchange:* The devices generate their public-private key pairs and send the public keys to each other. They also compute the Diffie-Hellman key.
3. *Authentication stage 1:* The protocol that is run at this stage depends on the association model. One of the goals of this stage is to ensure that there is no MITM in the communication between the devices. This is achieved by using a series of nonces, commitments to the nonces, and a final check of integrity checksums performed either through the OOB channel or with the help of the user.
4. *Authentication stage 2:* The devices complete the exchange of values (public keys and nonces) and verify their integrity.
5. *Link key calculation:* The parties compute the link key using their Bluetooth addresses, the previously exchanged values, and the Diffie-Hellman key constructed in phase 2.
6. *LMP authentication and encryption:* Encryption keys are generated in this phase, which is the same as the final steps of pairing in Bluetooth versions up to 2.0+EDR.

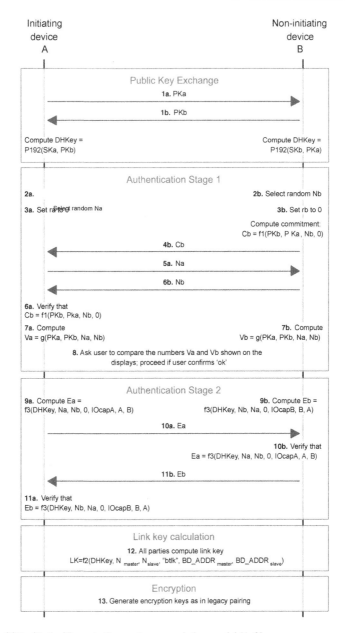

Fig. 2.4 SSP with the Numeric Comparison association model [1, 2]

The contents of messages sent during the SSP phase are outlined in Fig. 2.4 and the notations used are explained in Table 2.2.

Table 2.2 SSP protocol notation [1, 2]

Term	Definition
PKx	Public key of device X
SKx	Private key of device X
DHKey	Diffie-Hellman key generated after key exchange
Nx	Nonce generated by device X
rx	Random number generated by device X; equals 0 in the Numeric Comparison association model
Cx	Commitment value from device X
f1	One-way function used to compute commitment values
f2	One-way function used to compute the link key
f3	One-way function used to compute check values
g	One-way function used to compute numeric check values
IOcapX	Input/Output capabilities of device X
BD_ADDR	48-bit Bluetooth device address

Even though SSP improves the security of Bluetooth pairing, it has been shown that MITM attacks against Bluetooth 2.1+EDR, 3.0+HS, and 4.0 devices are possible by forcing victim devices to use the Just Works association model [2, 19–23]. Moreover, at least one of the proposed MITM attacks against Bluetooth SSP has already been implemented and mounted in practice [24]. Thus, the security of SSP should be further improved.

Chapter 3
Reasons for Bluetooth Network Vulnerabilities

Overall security in Bluetooth networks is based on *the security of the Bluetooth medium, the security of Bluetooth protocols*, and *the security parameters used in Bluetooth communication*. There are several weaknesses in the Bluetooth medium, Bluetooth protocols, and Bluetooth security parameters, which can significantly weaken the overall security of Bluetooth networks.

Section 3.1 discusses Bluetooth network vulnerability to eavesdropping. The weaknesses in the encryption mechanisms are explained in Sect. 3.2. Section 3.3 discusses weaknesses in PIN code selection. The weaknesses in association models of SSP are explained in Sect. 3.4. Section 3.5 discusses the weaknesses in Bluetooth device configuration.

3.1 Vulnerability to Eavesdropping

Because Bluetooth is a wireless RF communication system using mainly omnidirectional antennas, an eavesdropper is often not detected. The eavesdropper and eavesdropping equipment can be very far away from the communicating devices. Unencrypted transmissions are always easy prey for eavesdroppers. All the contents of unencrypted transmissions can be seen clearly. Even if Bluetooth data encryption (see Fig. 2.3 in Chap. 2) is used, all intercepted packets can be recorded for later cryptographical analysis, so the length of K_C should be as long as possible, i.e., the maximum K_C length of 128 bits should be used when possible [2].

There are three Bluetooth device classes: *class 1*, *class 2*, and *class 3*. The maximum transmit powers for class 1, class 2, and class 3 devices are 100 mW (20 dBm), 2.5 mW (4 dBm), and 1 mW (0 dBm) respectively. According to the Bluetooth specification [1], the reference sensitivity level of a Bluetooth device has to be −70 dBm or better [1, 2].

The *range of Bluetooth devices* depends on the class of devices at both ends, the sensitivity levels at both ends, and the level of obstacles. The quantity n is the

K. Haataja et al., *Bluetooth Security Attacks*, SpringerBriefs in Computer Science, DOI: 10.1007/978-3-642-40646-1_3, © The Author(s) 2013

so-called Path Loss (PL) exponent that can be adjusted to account for the amount of clutter in the path between the transmitter and the receiver. The *level of obstacles* can be roughly divided into four categories [2, 12]:

1. *None:* A free space without clutter in the transmit-receive path (n = 2.0).
2. *Light:* A lightly cluttered path such as an office environment with moveable walls (n = 2.5).
3. *Moderate:* A moderately cluttered path such as an office environment with fixed walls (n = 3.0).
4. *Heavy:* A heavily cluttered path in which the density of the materials used in the building's construction is very high (n = 4.0).

The most common case is a moderate level of obstacles, which is why the Bluetooth specification promises that the range of a Bluetooth device is from 10 (class 3 device) to 100 m (class 1 device) indoors when the level of obstacles is moderate. For application, as well as from the security analysis point of view, it is interesting to know how long is the range of Bluetooth devices [2].

Several assumptions must be made before proceeding with the range calculations. These assumptions reflect the typical characteristics of Bluetooth devices and can be summarized as follows [2, 12]:

1. Let us assume that a Bluetooth transmitter's (TX) power is either 0 dBm (class 3 device) or 20 dBm (class 1 device). The most common TX power for Bluetooth devices is 0 dBm (1 mW).
2. Let us assume that a Bluetooth receiver's (RX) sensitivity level is either −70 dBm (standard sensitivity level) or −80 dBm (enhanced sensitivity level).
3. Let us assume that Bluetooth transmit and receive antennas each have a gain of 0 dBi (decibels relative to an isotropic source).

Table 3.1 presents the *range of Bluetooth devices* with these prerequisites by using the formula $d = 10^{(PL-40)/(10n)}$, where d is the range of the Bluetooth devices, PL is the Path Loss value, and n is the PL exponent. A more precise definition of the range calculations can be found in [2, 12].

The *range of vulnerability to eavesdropping* can be calculated using the formula $d_e = 10^{(P_t+G_r+40)/(10n)}$, where d_e is the range of vulnerability to eavesdropping, P_t is the transmit power of the target piconet, G_r is the eavesdropper's antenna gain, and n is the PL exponent. A more precise definition of the range of vulnerability calculations can be found in [2, 12].

Several assumptions must be made before proceeding with the range of vulnerability calculations. These assumptions concern the typical physical properties of Bluetooth piconets. The calculations presented in Table 3.2 are made based on the following standard assumptions [2, 12]:

1. The target piconet devices are using omnidirectional antennas with a 0 dBi gain.
2. The level of obstacles is either light (n = 2.5) or moderate (n = 3.0) and there is 20 dB of additional attenuation on the signal as it exits the building.
3. The sensitivity level of the eavesdropper's radio is −100 dBm (enhanced sensitivity level).

Table 3.1 The range of Bluetooth devices [2, 12]

Level of obstacles:	n:	TX power (dBm):	RX sensitivity (dBm):	PL:	Range (m):
None	2.0	0	−70	70	32
None	2.0	0	−80	80	100
None	2.0	20	−70	90	316
None	2.0	20	−80	100	1000
Light	2.5	0	−70	70	16
Light	2.5	0	−80	80	40
Light	2.5	20	−70	90	100
Light	2.5	20	−80	100	251
Moderate	3.0	0	−70	70	*10*
Moderate	3.0	0	−80	80	*22*
Moderate	3.0	20	−70	90	*46*
Moderate	3.0	20	−80	100	*100*
Heavy	4.0	0	−70	70	6
Heavy	4.0	0	−80	80	10
Heavy	4.0	20	−70	90	18
Heavy	4.0	20	−80	100	32

Table 3.2 The range of vulnerability to eavesdropping [2, 12]

Level of obstacles:	n:	Target piconet TX power (dBm):	Eavesdropper's antenna gain (dBi):	Range of vulnerability for eavesdropping (m):
Light	2.5	0	0	40
Light	2.5	20	0	251
Light	2.5	0	20	251
Light	2.5	20	20	1585
Moderate	3.0	0	0	22
Moderate	3.0	20	0	100
Moderate	3.0	0	20	100
Moderate	3.0	20	20	464

4. The eavesdropper has two antennas, one with a gain of 0 dBi (omnidirectional) and another with a gain of 20 dBi (directional).
5. The target piconet devices use a transmit power of either 0 dBm (class 3 device) or 20 dBm (class 1 device).

As can be seen from Table 3.2, it would be easy, for example for an eavesdropper to park a car in a parking lot at a reasonable distance (up to 1585 m) from the target company's Bluetooth network, and start to intercept all Bluetooth activity within the range of vulnerability by using a laptop and a Bluetooth protocol analyzer [2].

Consider the following attack scenario. If the actual user data (the payload of a Bluetooth Baseband packet) is sent unencrypted, all the contents of the Baseband

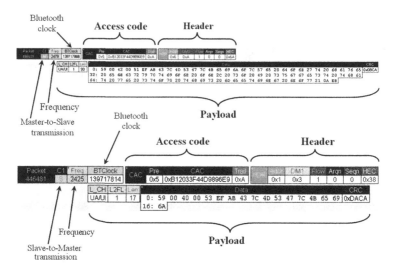

Fig. 3.1 An example of packet interception with a Bluetooth protocol analyzer when Bluetooth encryption is not used [2]

packet (an access code, a header, and the payload) and other relevant information (packet sequence number, direction of the transmission, frequency information, and Bluetooth clock information) can be displayed on the screen of a laptop by using a Bluetooth protocol analyzer, for example, as Fig. 3.1 illustrates [2].

The access code and the header are always sent unencrypted, so even when encryption is used an eavesdropper can always see the general piconet information, such as the piconet address of the active slave and the Baseband packet type used, from all of the packets. Based on this information, the eavesdropper may be able to figure out the authorization levels of the legitimate piconet devices, i.e., which Bluetooth device has access to a certain sensitive file. If the physical protection of the Bluetooth device is insufficient, the intruder can steal it and use it to obtain the desired sensitive file [2].

Bluetooth authenticates devices, not users, so this is also very important to keep in mind. The eavesdropper can also easily see whether the payload is encrypted or not. This can be seen directly from the Cyclic Redundancy Check (CRC) field. In this example (see Fig. 3.1), the received CRC field matches the CRC checksum calculated from the received user data, i.e., the payload is unencrypted. If the CRC field does not match the data, the protocol analyzer displays the CRC field, for example, in red indicating that the payload is encrypted [2].

3.2 Weaknesses in Encryption Mechanisms

As described in Chap. 2, Bluetooth encryption has many strengths, but it also has a few weaknesses. Perhaps the most significant weakness occurs when 128-bit encryption cannot be used. When two Bluetooth devices negotiate the parameters for encryption, the length of the encryption key is restricted by the Bluetooth device that has the shorter maximum encryption key length. For example, if one Bluetooth device can support only a 32-bit encryption key, the other Bluetooth device has to adjust to the situation and also use a 32-bit encryption key—otherwise encryption cannot be used at all [2].

Let us assume that an eavesdropper wants to use a brute-force method to find out a correct encryption key. This can be done by intercepting several encrypted Baseband packets and trying to decrypt them using a keystream generated by different experimental keys. Even though the keystream generator function has the piconet master's BD_ADDR and clock information as additional inputs, both of these values are supposed to be known by the eavesdropper. The eavesdropper can mark every incoming packet with its associated clock information before storing the packet, and if a packet contains any plaintext, the correct keystream will reveal it. The eavesdropper can use, for example, a dictionary in his key searching algorithm to focus the search on strings containing real words. An eavesdropper can also calculate a 16-bit CRC checksum from the payload, and check whether it matches the received CRC field. If the CRC checksums match, then decryption has been successful (if there are no bit errors in payload caused by the packet transfer via the RF link). Almost all commercially available Bluetooth protocol analyzers can automatically check whether the CRC checksums match, so this is not a problem for the eavesdropper [2].

Table 3.3 illustrates encryption weaknesses. The average search time, when using a naive guess-and-try *brute-force method*, measured in seconds is $2^{L-1}/n$, where L is the length of the encryption key in bits and n is the number of key search trials per second. Let us assume that an eavesdropper can make 2^{20} key search trials per second ($n = 2^{20}$) on the processing power of one computer, or 2^{40} key search trials per second ($n = 2^{40}$), for example, by using parasitic computing over the Internet. Further, let us assume that the eavesdropper's key searching algorithm can in all cases realize when the correct encryption key has been found. If the average search time is determined by 2^{20} trials per second, an adequate level of security can be achieved by using a 64-bit or longer encryption key. If the average search time is determined by 2^{40} trials per second, an adequate level of security can be achieved by using at least an 80-bit encryption key [2, 11, 12].

Let us assume that there are four slave devices (slaves A, B, C, and D) and one master device in a Bluetooth piconet. Slaves A, B, C, and D want to use 32-bit, 64-bit, 128-bit, and 128-bit encryption with the master, respectively. As can be seen from Table 3.3, slave A has only primitive protection against eavesdroppers. In addition, an eavesdropper may be able (at least in theory) to decrypt the packets of slave B in a reasonable time, because it takes only an average of 97 days assuming that the average search time is determined by 2^{40} trials per second. All this can

Table 3.3 Encryption weaknesses [2, 11, 12]

K_C length (bits):	Average search time at 2^{20} trials per second:	Average search time at 2^{40} trials per second:
8	≈ 122 microseconds	≈ 116 picoseconds
16	≈ 31 microseconds	≈ 30 nanoseconds
24	≈ 8 seconds	≈ 8 microseconds
32	≈ 34 minutes	≈ 2 milliseconds
40	≈ 6 days	≈ 500 milliseconds
48	≈ 4 years	≈ 128 seconds
56	≈ 1090 years	≈ 9 hours
64	≈ 278922 years	≈ 97 days
72	≈ 71 million years	≈ 68 years
80	≈ 18 billion years	≈ 17433 years
128	$\approx 5.1 \times 10^{24}$ years	$\approx 4.9 \times 10^{18}$ years

be achieved with a naive guess-and-try method. A more sophisticated method for the eavesdropper would be exploiting the results described in [8] (see Sect. 4.1), in which the improved method yields approximately 30 % faster decryption times: the guess-and-try method's 97 days versus the more sophisticated method's 68 days. Moreover, if slaves A and B are exchanging sensitive files with slaves C and D, the better security of slaves C and D, i.e., the protection of 128-bit encryption, is lost. Therefore, if security is very important, the master should not accept encryption key lengths shorter than 128 bits [2].

3.3 Weaknesses in PIN Code Selection

The weakest point in the security chain of events (see Fig. 2.1 in Chap. 2) of Bluetooth versions up to 2.0+EDR is the first phase in which a user selects a PIN code. The PIN code can be as long as 128 bits (16 bytes), so it can contain up to sixteen 8-bit characters. However, long PIN codes are quite hard to remember, so users usually use only four digits. This makes an eavesdropper's work much easier, because she needs to go through only 10000 possible PIN values and witness the initial pairing process between the target devices in order to get all the required information for various attacks (see Sects. 4.1–4.4). It is worth noting that the attacker needs an average of only 5000 PIN guesses to find out the correct value when a four-digit PIN code is used. On the other hand, if the user decides to use sixteen 8-bit characters, it is very likely that she will write down the PIN code on a piece of paper. This is another weak point, because this piece of paper must be kept secret [2].

A 16-digit PIN code composed of the characters 0,...,9 achieves $16 \times \log_2 10 \approx 53$ bits of entropy, while a PIN code of 16 case-sensitive alphanumerical characters yields $16 \times \log_2 62 \approx 95$ bits of entropy when a 62-character set is used, i.e., the

62-character set consists of the characters 0,...,9, 'a',...,'z' and 'A',...,'Z'. Therefore, a Bluetooth PIN code that achieves 128 bits of entropy can be provided by using sixteen 8-bit characters, i.e., a 256-character set is used ($16 \times \log_2 256 = 128$ bits of entropy). For example, the extended American Standard Code for Information Interchange (ASCII) character set has 256 characters ($2^8 = 256$), so there are eight bits of entropy for each 8-bit character (byte) [2].

Two devices become *paired* when they start communicating with the same PIN code, generate the same link key, and then use the link key for authenticating at least the current communication session (see Figs. 2.1 and 2.2 in Chap. 2). When devices are paired, they can either store their link keys for use in subsequent authentications, or discard them and repeat the pairing process each time they connect. If the link keys are stored, the devices are *bonded*. Users that use bonded Bluetooth devices do not have to remember long PIN codes. On the other hand, these bonded devices can be a security risk if the physical protection of the devices is insufficient [2].

Many different kinds of Bluetooth devices, such as headsets, keyboards, and printer adapters, have very short fixed PIN codes, often only four digits long. This is clearly a big security risk, so Bluetooth device manufacturers should take security issues more seriously. We strongly recommend that sixteen 8-bit character PIN codes should be used when possible. However, in the case of a limited User Interface, it may not be possible to use the 256-character set to provide a PIN code that achieves 128 bits of entropy. Moreover, user understanding of security issues is very important for protecting sensitive data against eavesdroppers and hackers. Many users have no idea how to configure their Bluetooth devices' security settings correctly [2].

3.4 Weaknesses in Association Models of SSP

As described in Chap. 2, Bluetooth 2.1+EDR, 3.0+HS, and 4.0 specifications support SSP to improve the security of pairing by providing protection against passive eavesdropping and MITM attacks. SSP uses four association models: *OOB*, *Numeric Comparison*, *Passkey Entry*, and *Just Works* (see Chap. 2). The choice of association model depends on the device's IO capabilities (see Table 2.1 in Chap. 2). Perhaps the most significant weakness occurs when at least one of the devices has neither input nor output capability and an OOB cannot be used. In this case, the Just Works association model is used, in which the user is simply asked to accept the connection. Therefore, the Just Works association model clearly provides no MITM protection [2].

SSP consists of six phases: *Capabilities exchange*, *Public key exchange*, *Authentication stage 1*, *Authentication stage 2*, *Link key calculation*, and *LMP authentication and encryption* (see Chap. 2). The last phase in SSP is the same as the final steps of pairing in Bluetooth versions up to 2.0+EDR (see Figs. 2.1 and 2.3 in Chap. 2) [2].

Just as in the Bluetooth versions up to 2.0+EDR, the weakest point in an SSP-enabled Bluetooth device's security chain of events is the first phase. Instead of selecting a PIN code, SSP-enabled devices will exchange their IO capabilities (see Table 2.1 in Chap. 2) to determine the proper association model to be used. If an

OOB, Numeric Comparison, or Passkey Entry association model is used, MITM protection will be automatically provided. However, it has been shown that MITM attacks against Bluetooth 2.1+EDR/3.0+HS/4.0 devices are possible by forcing the victim devices to use the Just Works association model [2, 9–23] (see Chap. 5). Moreover, at least one of the proposed MITM attacks against Bluetooth SSP has already been implemented and mounted in practice [24]. By far the best way to prevent MITM attacks is to use NFC as an OOB channel [2].

3.5 Weaknesses in Device Configuration

The default settings of Bluetooth devices usually provide no security at all: the device is set as discoverable (i.e., public security level) and nonsecure (i.e., non-secure security mode). Therefore, an attacker can discover the BD_ADDR of the target device in a few seconds and perform various attacks (see Sects. 4.1–4.4) against it [2].

It is worth noting that Bluetooth is rarely switched on by default, so a user has to switch it on from the device's settings before any Bluetooth attacks against that device are possible. Moreover, many users want to save the batteries of their Bluetooth devices so Bluetooth is often switched off when there is no need to use it for a long time [2].

It is very important that users know how to configure their Bluetooth devices correctly to achieve the best available level of security. In addition, Bluetooth device manufacturers should implement their Bluetooth devices in a more secure way using default factory settings: for example, if a device uses a fixed PIN code, it should be as long as possible and also as hard as possible to guess. If both communicating devices support SSP and NFC, NFC should always be used as an OOB channel (see Chap. 2). Moreover, application layer key exchange and encryption methods can be used as extra security in addition to the Bluetooth built-in security [2].

Chapter 4
Comparative Analysis of Bluetooth Security Attacks

As an interconnection technology, Bluetooth has to address all the traditional security problems, well known from distributed networks [25]. In addition, security issues in wireless ad hoc networks are much more complex than those of more traditional wired or centralized wireless networks. Moreover, Bluetooth networks are formed by radio links, which means that there are additional security aspects whose impact is not yet well understood [2].

Since there are now (and will be in the near future) billions of Bluetooth devices in use without SSP's improved security features (see Chap. 2), malicious security violations are not expected to decrease in the near future. On the contrary, these old Bluetooth devices will be sold for many years to come, thus making security concerns even more alarming. Moreover, attacks against SSP are also possible as we discussed in Chap. 2 and Sect. 3.4. Therefore, the Bluetooth security architecture needs to be further upgraded to meet these new threats [2].

Security threats in distributed networks (such as Bluetooth) can be divided into three categories: disclosure threat, integrity threat, and Denial-of-Service (DoS) threat. *Disclosure threat* means that information can leak from the target system to an eavesdropper that is not authorized to access the information. *Integrity threat* concerns the deliberate alteration of information in an attempt to mislead the recipient. *DoS threat* involves blocking access to a service, either making it unavailable or severely limiting its availability to an authorized user [2, 12].

Bluetooth security is currently a very active research area, because Bluetooth devices are widely used all over the world. In addition to our Bluetooth research, there are several research papers, research reports, and homepages about the security vulnerabilities of Bluetooth, i.e., descriptions of many different kinds of disclosure, integrity, and DoS attacks. Disclosure and integrity attacks typically compromise some sensitive information and therefore can be very dangerous, while DoS attacks typically only annoy Bluetooth network users and are considered to be less dangerous [2].

Powerful directional antennas can be used to increase the scanning, eavesdropping, and attacking range of almost any kind of Bluetooth attack considerably. One

K. Haataja et al., *Bluetooth Security Attacks*, SpringerBriefs in Computer Science,
DOI: 10.1007/978-3-642-40646-1_4, © The Author(s) 2013

very good example of a long-distance attacking tool is the BlueSniper Rifle [26, 27]. This is a rifle stock with a powerful directional antenna attached to a small Bluetooth-compatible computer. Scanning, eavesdropping, and attacking can be done over a mile away from the target devices. Moreover, anyone with some basic skills and a few hundred dollars can build her own BlueSniper Rifle. Therefore, the possibility that an attacker is using range enhancement to improve the performance of the attacks should be taken seriously [2, 26, 27].

Nowadays it is also possible to transform a standard $30 Bluetooth dongle into a full-blown Bluetooth sniffer [28]. We have also verified this fact in our research laboratory [2] with many different Cambridge Silicon Radio (CSR)-based Bluetooth Universal Serial Bus (USB) dongles supporting Bluetooth versions up to 2.0+EDR (Enhanced Data Rate). In addition, tools for reverse engineering the firmware of CSR-based Bluetooth dongles are available [29]. The tools include a disassembler for the official firmware and an assembler that can be used for writing custom firmware. With these tools anyone can now write custom firmware for CSR-based Bluetooth dongles to include raw access for Bluetooth sniffing. The tools also include the source code for sniffing Bluetooth under Linux. Moreover, it is expected that in the near future techniques for finding hidden (non-discoverable) Bluetooth devices in an average of one minute [30, 31] will be ported onto a standard CSR dongle via custom firmware [2]. This will open new doors for practical Bluetooth security research and it will also provide a cheap basic weapon to all attackers for Bluetooth sniffing. It is expected that Bluetooth sniffing will soon become a very popular sport among attackers and hackers, thus making Bluetooth security concerns even more alarming [2, 28-32].

Sections 4.1–4.3 explain some typical disclosure, integrity, and DoS threats, respectively. Some typical threats which cannot be classified as only one single threat (so-called *multithreats*) are explained in Sect. 4.4. Besides explaining the existing Bluetooth security analysis tools implemented by other researchers, we also explain our own Bluetooth security analysis tools: the scripts and/or source codes of our security analysis tools exist, but they will not be released in any public domain because they can be very dangerous due to their efficiency.

4.1 Disclosure Threats

A *BlueSnarfing attack* [33, 34] (also referred to as a *BlueStumbling attack*) means that an attacker connects to the target device without alerting its owner and steals some sensitive information, such as an entire phonebook, calendar notes, or text messages. At least three BlueSnarfing applications exist: Adam Laurie's *BlueSnarf* [34], Ollie Whitehouse's *RedSnarf* [35], and Bluediving Project's *Bluediving* [36]. The source codes and binaries of BlueSnarf and RedSnarf have not been released, while the source codes and binaries of Bluediving have been made public. Bluediving runs on Linux and is based on the BlueZ [37] protocol stack. *BlueZ* is the official Bluetooth protocol stack for Linux environments and it is included in the Linux 2.4 (or later) kernel series [2, 33-37].

The Object Exchange Protocol (OBEX) can be used to exchange business cards between Bluetooth devices using the OBEX protocol's object push feature, i.e., pushing (sending) business cards (objects) to Bluetooth devices using the OBEX protocol's Put operation. In most cases, this service does not require authentication. BlueSnarfing is based on the exploitation of the OBEX protocol's object pull feature instead of the object push feature. This feature is used for pulling (receiving) objects from Bluetooth devices using the OBEX protocol's Get operation. A BlueSnarfing attack conducts an OBEX Get request for known filenames such as telecom/pb.vcf (the phonebook file) or telecom/cal.vcs (the calendar file). A detailed description of the OBEX protocol and its operations can be found in [38], while a detailed description on how to implement a BlueSnarfing attack in practice can be found in [2, 36, 38].

The success of a BlueSnarfing attack depends very much on the vendor's implementation of the Bluetooth protocol stack for the target device. Therefore, the attack works only if the protocol stack of the target device is poorly implemented, i.e., there are serious flaws in the authentication and data transfer mechanisms of some Bluetooth devices. A list of the devices known to be vulnerable to a BlueSnarfing attack without firmware/software update can be found in [34]. Moreover, BlueSnarfing is normally only possible and dangerous if the target device's security level is set to public, i.e., the target device is discoverable, but there are also ways to find non-discoverable devices, for example, by using brute-force scanning, a Bluetooth protocol analyzer, 79 receivers in parallel, or the techniques introduced in [2, 30, 34].

An *Off-Line PIN Recovery attack* [13, 35] (also referred to as an *Off-Line PIN Crunching attack*) is based on intercepting the IN_RAND value, LK_RAND values, AU_RAND value, and SRES value, and after that trying to calculate the correct SRES value by guessing different PIN values until the calculated SRES equals the intercepted SRES (see Figs. 2.1 and 2.2 in Chap. 2). It is worth noting that SRES is only 32 bits long. Therefore, an SRES match does not necessarily guarantee that an attacker has discovered the correct PIN code, but the chances are quite high especially if the PIN code is short [2, 12, 13, 35, 39].

An Off-Line PIN Recovery attack is dangerous only if the PIN code is short and it has no case-sensitive alphanumerical characters (and perhaps some other characters as well). Moreover, an attacker must intercept the traffic of the initial pairing process between two Bluetooth devices when they meet for the first time (see Chap. 2). This can be arranged in many different ways, for example, by sending a Bluetooth device anonymously to the target person as a prize in some competition. Another possible way to witness the initial pairing process is to disrupt the connection establishment process between two devices, for example, by disrupting the Physical Layer (PHY) in such a way that the user thinks something is wrong and deletes previously stored link keys. After that the user initiates a new pairing process and the attacker can intercept all the required inputs for an Off-Line PIN Recovery attack [2].

An *Enhanced implementation of Off-Line PIN Recovery attack* [8] is an average of 30% faster than the original Off-Line PIN Recovery attack described in [13, 35]. It is based on the optimization of SAFER+ [6] (see Chap. 2) using the algebraic manipulation of the SAFER+ round. Other more recent results on the optimization of SAFER+ rounds can be found, for example, in [2, 6, 8, 9].

Three methods which can *force two target devices to repeat the initial pairing process* are also proposed in [8]:

1. The first method (and two other methods as well) is based on the fact that Bluetooth specifications allow Bluetooth devices to forget a link key: when the master sends AU_RAND to the slave during the authentication, the slave sends an "LMP not accepted" message in return to let the master know that it has forgotten the link key, i.e., the slave does not send the SRES value as in normal authentication. Therefore, the master is convinced that the slave has lost the link key and the pairing process is restarted.
2. The second method works in the following way: at the beginning of the authentication process, the master is supposed to send AU_RAND to the slave. If an attacker sends IN_RAND to the slave before the master sends AU_RAND, the slave device is convinced that the master has lost the link key and pairing is restarted.
3. The third method works in the following way: when the master sends AU_ RAND to the slave during authentication, the attacker sends a random SRES message to the master, causing the authentication process to restart, and the same kind of repeated attempts will be made. After a certain number of failed authentication attempts, the master is expected to declare that the authentication process has failed and the pairing process is restarted. The required number of failed authentication attempts is implementation-dependent.

These three methods require that the attacker has a custom Bluetooth device, such as a protocol analyzer, for cloning the BD_ADDR values of the target devices, i.e., the methods are based on the impersonation of target devices. Moreover, the Bluetooth user is required to enter a PIN code again during the new pairing process, and therefore a suspicious user may realize that her device is under attack [2, 8].

An *Off-Line Encryption Key Recovery attack* [35] extends an Off-Line PIN Recovery attack [8, 13, 35] and it is based on intercepting the IN_RAND value, LK_RAND values, AU_RAND value, SRES value, and EN_RAND value, i.e., it requires that the attacker intercepts all the values required for the Off-Line PIN Recovery attack and also the EN_RAND value. When the PIN code is discovered via an Off-Line PIN Recovery attack, the attacker can produce the required ACO using the $E_1(AU\_RAND_A, BD\_ADDR_B, Link\ key)$ function, where the Link key is K_A or K_{AB}. After that the attacker can recover the encryption key (see Fig. 2.3, in Chap. 2) using the formula $K_C = E_3(EN\_RAND_A, ACO, Link\ key)$, where the Link key is K_A or K_{AB}. Therefore, an Off-Line Encryption Key Recovery attack is dangerous only when an Off-Line PIN Recovery attack or its enhanced implementation has been completed successfully [2, 35].

A *Brute-Force BD_ADDR Scanning attack* [2, 35] uses a brute-force method only on the last three bytes of a BD_ADDR, because the first three bytes are publicly known and can be set as fixed. A Brute-Force BD_ADDR Scanning attack is perhaps most feasible when the target devices are Bluetooth mobile phones, because millions of vulnerable Bluetooth mobile phones are used every day all over the world. *RedFang* [40] is a security analysis tool for finding non-discoverable Bluetooth devices by brute-forcing the last three bytes of BD_ADDR and doing a name inquiry. It runs on Linux and requires a BlueZ protocol stack [37] and at least one Bluetooth USB dongle to work [2, 35, 40].

We designed, implemented, and tested our own tool to carry out this attack. Our *Brute-Force BD_ADDR Scanning Security Analysis Tool [2, 41] is on average four times faster than RedFang*, because it runs on special hardware, LeCroy's Bluetooth protocol analyzer [42], which can use Bluetooth radio much more efficiently than a normal PC with a Bluetooth USB dongle [2, 41].

We successfully performed a Brute-Force BD_ADDR Scanning attack using LeCroy's Bluetooth protocol analyzer [42] with one Bluetooth 1.1-compatible radio unit and an unmodified Bluetooth 1.1-compatible Nokia 6310i mobile phone [43]. LeCroy BTTracer/Trainer v2.2 software [42], which provides CATC Scripting Language [44], was also used in our practical experiment [2, 41].

We used CATC Scripting Language to create our *Brute-Force BD_ADDR Scanning Script*, which works in the following way [2, 41]:

1. Set the scanning area.
2. Set remote BD_ADDR for the next BD_ADDR trial, i.e., set a BD_ADDR value for the next connection attempt.
3. Try to create a basic ACL link between the protocol analyzer and a remote device by using the BD_ADDR value set in step 2. If the connection attempt fails, go back to step 2. Otherwise, the Brute-Force BD_ADDR Scanning Script has found a non-discoverable device (see Chap. 2).
4. Perform a remote name inquiry and a disconnection with the target device. If there is more scanning left to do, go back to step 2.

All scanning information can be stored in a logfile for later analysis. Figure 4.1 illustrates an example of a successful Brute-Force BD_ADDR Scanning attack using our Brute-Force BD_ADDR Scanning Script. The remote BD_ADDR value is changed to a new value (see rows 1, 4, 7, and 17) after every connection attempt if there is more scanning left to do. Two failed connection attempts are performed (see rows 2–3 and 5–6) before the successful connection establishment (see rows 8–10) in which a remote name inquiry (see rows 11–13) and disconnection (see rows 14–16) with the target device is also performed. Now an attacker has discovered the BD_ADDR of the target device and further attacks (see Sects. 4.1–4.4) against that device can be performed. Because there is more scanning left to do, the remote BD_ADDR value is changed again to a new value (see row 17), and a new connection attempt is performed (see rows 18–19). In our practical experiment, the total time for scanning through 2000 BD_ADDRs was 174 min and 20 s. Hence, the average time required for one reliable BD_ADDR trial is 5.2 s ($10460 s/2000 \approx 5.2 s$). The average scanning time for all 8388608 BD_ADDR trials (i.e., a 24-bit address space gives 16777216 different BD_ADDR values and an attacker needs on average 8388608 BD_ADDR guesses to find out the correct value) with Brute-Force BD_ADDR Scanning Script is 1.4 years ($8388608 \times 5.23 s \approx 43872420 s \approx 1.4 years$) when using only one radio unit. A second radio unit or another LeCroy Bluetooth protocol analyzer can be used to speed up the scanning process. If, for example, 25

```
(1)   Remote BD_ADDR for this trial is: 0002eeb0294b
(2)   HCI_Evt> Connection_Error
(3)     Error                     : Page Timeout
(4)   Remote BD_ADDR for this trial is: 0002eeb0294c
(5)   HCI_Evt> Connection_Error
(6)     Error                     : Page Timeout
(7)   Remote BD_ADDR for this trial is: 0002eeb0294d
(8)   HCI_Evt> Connection_Complete
(9)     BD_ADDR                   : 0002EEB0294D
(10)    HCI Handle                : 0x0004
(11) HCI_Evt> Remote_Name_Request_Complete
(12)    BD_ADDR : 0002EEB0294D
(13)    Name      : "Nokia 6310i"
(14) HCI_Evt> Disconnection_Complete
(15)    BD_ADDR                   : 0002EEB0294D
(16)    Reason                    : No Connection
(17) Remote BD_ADDR for this trial is: 0002eeb0294e
(18) HCI_Evt> Connection_Error
(19)    Error                     : Page Timeout
```

Fig. 4.1 An example of a successful Brute-Force BD_ADDR Scanning attack [2, 41]

compact size LeCroy Merlin II [42] protocol analyzers are used for a Brute-Force BD_ADDR Scanning attack with our Brute-Force BD_ADDR Scanning Script, it takes on average 20.3 days ($43872419.84 \, s/25 \approx 1754897 \, s \approx 20.3 \, days$) [2, 41].

For comparison, RedFang needs as many as 100 concurrent Bluetooth USB dongles to achieve the same result, and it is very likely that due to greater RF interference they will not work as reliably as 25 concurrent protocol analyzers, i.e., there are only 79 different Baseband frequencies and therefore 100 concurrent Bluetooth USB dongles within the range will cause RF interference. Moreover, an attacker has to use a laptop or laptops with several USB hubs to make this kind of attack feasible. Brute-Force BD_ADDR Scanning attacks can be feasible and dangerous if an attacker has enough resources, namely equipment, money, time, and will, and good software tools such as Brute-Force BD_ADDR Scanning Script or RedFang. In addition, the attacker has to know the manufacturer of the target device, because brute-forcing a 48-bit address space is not feasible. Moreover, just discovering the BD_ADDR of the target device will not give access to any sensitive files. Therefore, some additional attacks must also be performed and the target device must be somehow vulnerable to them [2, 41].

A Brute-Force BD_ADDR Scanning attack is perhaps most feasible when the target devices are Bluetooth mobile phones, since Nokia is still the world's leading mobile phone manufacturer. Another good guess for the mobile phone's manufacturer is Samsung, which is the world's second largest mobile phone manufacturer. Moreover, millions of vulnerable Bluetooth mobile phones [33, 34, 45–48] are used every day all over the world [2, 41].

An attacker can also try to use a virus that turns all infected Bluetooth-enabled PCs and laptops into scanning devices to speed up the Brute-Force BD_ADDR Scanning attack. Moreover, the attacker can speed up the search process and increase the probability of finding several Bluetooth mobile phones by first scanning for discoverable mobile phones. Based on the BD_ADDRs of the discoverable Bluetooth mobile phones, the attacker can determine the most commonly used company-assigned values (see Chap. 2) in this particular geographical area and scan the BD_ADDRs that are near those of the discoverable mobile phones first. It is very likely that the BD_ADDR values are almost the same within the same geographical area, because the process of assigning company-assigned values is not completely random. Several practical experiments in our Bluetooth security laboratory showed that when testing Brute-Force BD_ADDR Scanning even with very small scanning ranges, such as scanning only 100 or 200 BD_ADDRs that are near the BD_ADDR of our laboratory's Bluetooth mobile phone, typically several additional Bluetooth mobile phones were discovered accidentally [2, 41].

Besides a Brute-Force BD_ADDR Scanning attack, *techniques for finding hidden Bluetooth devices in an average of one minute* have been proposed [30] and even implemented [31]. Spill et al. also implemented an open-source Bluetooth sniffer [31] that operates on a single channel. The Universal Software Radio Peripheral (USRP) [49] was used as a radio device to eavesdrop on Bluetooth packets. This is the hardware device associated with the GNU Radio Project [50], which is developing an open-source framework for implementing software radio systems, i.e., systems in which radio devices are implemented in software [2, 30, 31, 49, 50].

Due to the buffering and asynchronous nature of the GNU Radio framework and the hardware restrictions of the USRP, no working prototype of the Bluetooth sniffer that supports frequency hopping has been implemented yet. However, the current version of the Bluetooth sniffer is still capable of finding hidden (non-discoverable) Bluetooth devices in the range of vulnerability in an average of one minute by using the following techniques [2, 30]:

1. *Phase 1*: The LAP (see Chap. 2) can be determined in a straightforward manner since it is present in every Baseband packet in the form of a constant 72-bit pattern called the access code: this contains the 24-bit LAP along with its 34-bit checksum and 14 bits of synchronization and error detection data. Therefore, the LAP can simply be read from an intercepted Baseband packet and validated by its checksum. In order to eavesdrop a Baseband packet, it is sufficient to stay tuned to a single channel and wait for a Baseband packet to fly by. Since the channel hopping rate is very high (1,600 hops/second), waiting for one second is more than enough in order to intercept a Baseband packet [2, 30].
2. *Phase 2*: Each Baseband packet has a 10-bit header with an 8-bit Header Error Check (HEC) field that is calculated from the UAP (see Chap. 2). Spill et al. noticed that it was possible to reverse the HEC in order to reveal the UAP in almost real-time.
3. *Phase 3*: Since the first byte of the NAP (see Chap. 2) is almost always zero in practice, the remaining byte can be brute-forced by sending at most 256 pings to all the possible remaining BD_ADDR combinations.

Pinging a Bluetooth device takes approximately one second. The devices also need to find one another and identify themselves. This takes up to a second, because both devices have unique hopping patterns and these patterns need to coincide on a frequency before communication can take place. Therefore, it takes only an average of 4.3 min (phase 1 + phase 2 + phase 3 ≈ 1 s + 0 s + 2 s × 256/2 ≈ 4.3 min) to find a hidden Bluetooth device in the range of vulnerability. Moreover, IEEE's OUI database [5] (see Chap. 2) can be used to make educated guesses regarding the last byte of the NAP rather than blindly brute-forcing it. Typically, filtering the OUI list for vendor prefixes yields only a few dozen brute-force candidates, thus further reducing the time requirement. Spill et al. performed a practical experiment [30, 31] in which they first revealed the first four bytes (LAP and UAP) of the hidden BD_ADDR (5B:00:FA:C2) via phase 1 and phase 2. In phase 3, they filtered the OUI list for vendor prefixes ending in 5B and got 41 brute-force candidates. In their practical experiment, it took at most 2 min to find a hidden Bluetooth device using the techniques defined in phase 1, phase 2, and phase 3. Therefore, it takes only an *average of one minute* to find a hidden Bluetooth device [2, 30, 31].

Every Bluetooth device has some characteristics which are unique (BD_ADDR), manufacturer specific (the first three bytes of BD_ADDR), and model specific (SDP records). Moreover, every Bluetooth device that offers services to other Bluetooth devices will announce its SDP records via the Service Discovery Protocol (SDP). Therefore, remote Bluetooth devices can query other Bluetooth devices based on the offered capabilities. SDP records, which consist of information on how to access the particular service, are returned to the querying device. Certain values of SDP records can be used to calculate a fingerprint value that is used to determine the device model and the firmware version of the target device [2].

A *BluePrinting attack* [51] is used to determine the manufacturer, device model, and firmware version of the target device. For example, an attacker can use Blueprinting to generate statistics about Bluetooth device manufacturers and models, and to find out whether there are devices in the range of vulnerability that have issues with Bluetooth security. *BluePrint* [52] is a tool for performing BluePrinting attacks. It runs on Linux and it is based on the BlueZ protocol stack [37]. BluePrinting attacks work only when the BD_ADDR of the target device is known [2, 51, 52].

Our practical experiment, *Interception of Packets attack* [2, 53], was conducted in order to demonstrate the importance of data encryption and to show how easy it is for an eavesdropper to intercept all packets exchanged via air. The equipment needed for our practical experiment was a laptop connected to the LeCroy BTTracer/Trainer protocol analyzer [42], LeCroy BTTracer/Trainer v2.2 software [42], our Bluetooth Chat Software [2, 53], and PCs with Bluetooth USB dongles. We created our own Internet Relay Chat (IRC) [54] style software that consists of BTChatd (a Bluetooth Chat server for Linux), BTChat (a Bluetooth Chat client for Linux), and BTChatJava (a Bluetooth Chat Java client for Linux and Windows) [2, 53]. Bluetooth Chat Software runs on Linux/Windows and it requires only normal PCs/laptops and Bluetooth USB dongles to work [2, 53].

All PCs had a Bluetooth USB adapter and Bluetooth protocol stack up and running. One PC was the piconet master running BTChatd [2, 53] over BlueZ [37] in Linux. The other seven PCs were piconet slaves running BTChat [2, 53] in Linux or BTChatJava [2, 53] in Linux/Windows. Figure 4.2 illustrates the startup of the BlueZ protocol stack [37] (see rows 1–3) when encryption is not used (i.e. a nonsecure secu-

```
 (1)  May 27 10:25:25 itmw-ope24 hcid[7092]: HCI daemon ver 2.4 started
 (2)  May 27 10:25:25 itmw-ope24 bluetooth: hcid startup succeeded
 (3)  May 27 10:25:25 itmw-ope24 bluetooth: sdpd startup succeeded
 (4)  May 27 10:25:31 itmw-ope24 btchatd: Serial Port service registered
 (5)  May 27 10:25:31 itmw-ope24 btchatd: Logging chat to /tmp/btlog.txt
 (6)  May 27 10:25:31 itmw-ope24 btchatd: BtCHATD, BlueZ RFCOMM chat server running!
 (7)  May 27 10:25:31 itmw-ope24 btchatd: Waiting for connection on RFCOMM channel 10
 (8)  May 27 10:25:31 itmw-ope24 btchatd: Tuomas Kepanen, kepanen@cs.uku.fi
 (9)  May 27 10:25:55 itmw-ope24 btchatd: Connection from [00:05:4E:00:68:64]
(10)  May 27 10:26:52 itmw-ope24 btchatd: Connection from [00:05:16:48:0C:F5]
(11)  May 27 10:27:07 itmw-ope24 btchatd: Connection from [00:05:16:48:12:4C]
(12)  May 27 10:27:32 itmw-ope24 btchatd: Connection from [00:05:16:48:0C:F2]
(13)  May 27 10:27:48 itmw-ope24 btchatd: Connection from [00:05:16:48:0C:F3]
(14)  May 27 10:28:02 itmw-ope24 btchatd: Connection from [00:05:16:48:10:58]
(15)  May 27 10:29:32 itmw-ope24 btchatd: Connection from [00:05:16:48:12:55]
```

Fig. 4.2 The startup of the BlueZ protocol stack when encryption is not used, the startup of BTChatd, and connection establishments of seven slaves [2, 53]

rity mode), startup of BTChatd [2, 53] (see rows 4–8), and connection establishments of seven slaves (see rows 9–15). The notation hcid means Host Controller Interface (HCI) daemon and sdpd means SDP daemon [2, 53].

BTChatd, BTChat, and BTChatJava use the Radio Frequency Communication (RFCOMM) protocol to emulate a standard serial port. The chat software is implemented on top of RFCOMM, because it is supported in every Bluetooth protocol stack, and applications over RFCOMM are easy to implement. Because the piconet master is in a nonsecure security mode, each of the slaves can establish a connection without authentication, authorization, and encryption, i.e., the link is unprotected. Figure 4.3 illustrates the chat session between the active piconet slaves using BTChatJava on Windows when encryption is not used. The purpose of the chat software is to enable IRC-style communication between a maximum of seven active piconet slaves via the piconet master, which only relays messages to all the connected slaves [2, 53].

An eavesdropper has to synchronize with the piconet master in order to intercept the packets exchanged via air. Commercially available Bluetooth protocol analyzers, such as LeCroy's Bluetooth protocol analyzer [42], usually require only a brief contact with the piconet master to extract the BD_ADDR of the master and hop sequence information. This can be done, for example, by using a general inquiry in which all devices in range return their Frequency Hop Synchronization (FHS) packets. Therefore, if the security level of the piconet master is set as public, the required synchronization with the master can be done very quickly [2, 53].

On the other hand, a serious eavesdropper does not want to transmit anything that might disclose her location or intentions. It is also possible to synchronize with the piconet's hop sequence without transmitting at all. It can be done, for example, in the following way. The eavesdropper listens to one of the 32 inquiry hop frequencies in order to detect an inquiry. When one is detected, the eavesdropper's radio begins to hop along with the inquirer, checks each response frequency, and records the FHS packet of each responder. In this way, over a period of time, the eavesdropper will discover the identities of the Bluetooth devices that are within the range of vulnerability [2, 53].

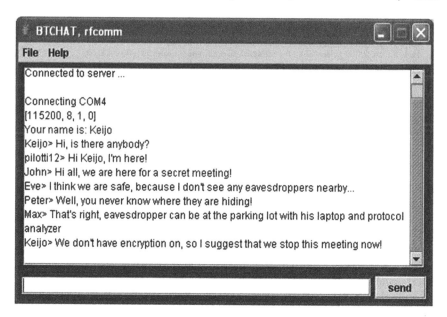

Fig. 4.3 The chat session between the active piconet slaves when encryption is not used [2, 53]

More difficult for an eavesdropper is the situation where all the target piconet devices are configured as non-discoverable devices, because no inquiries will occur in the range of vulnerability, i.e., the security level is set as private for each device. In this case, the eavesdropper has to monitor traffic on various hop frequencies perhaps over a long period of time in order to intercept a page or FHS packet. A page packet informs the eavesdropper that a page process, including the transmission of an FHS packet, is taking place. It is also possible to use several receivers in parallel to increase the probability of intercepting all useful information: for example, 79 receivers will intercept all Bluetooth activity within the range of vulnerability. Moreover, techniques for finding hidden Bluetooth devices in an average of one minute have been proposed [30] and even implemented [2, 31, 53].

Figure 4.4 illustrates the results of an Interception of Packets attack when encryption is not used. It is very clear that an eavesdropper can easily understand the contents of the intercepted data and save it, for example, to a text file for later use. The same message is shown several times on the screen of the eavesdropper's laptop, because the piconet master relays all messages to all participants of the chat session, i.e., the eavesdropper has to synchronize only with the piconet master in order to eavesdrop on a chat session between all active piconet slaves. The 16-bit CRC field calculated from the Baseband packet payload also matches the received CRC field, because our Bluetooth protocol analyzer displays the CRC field in white, i.e., white in the CRC field is used to indicate the match between the CRC checksums, while red indicates a mismatch between the CRC checksums. Now the eavesdropper knows for sure that the messages are being sent via a nonsecure link [2, 53].

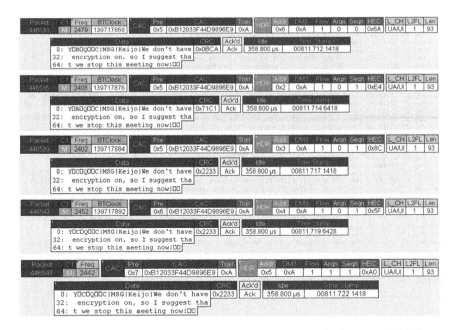

Fig. 4.4 The results of an Interception of Packets attack when encryption is not used [2, 53]

Figure 4.5 illustrates the startup of the BlueZ protocol stack (see rows 1–5) when encryption is used (Bluetooth security mode 2 is used), the startup of BTChatd (see rows 6–10), and the connection establishments of the piconet slaves (see rows 11–30, where the notation sba means the BD_ADDR of the source device, i.e., the BD_ADDR of the master device, and dba means the BD_ADDR of the destination device) [2, 53].

The startup of the BlueZ protocol stack illustrated in rows 1–5 is similar to that in Fig. 4.2 except that now a Bluetooth security manager (see row 2) is also needed, i.e., if any part of Bluetooth security takes place automatically, a security manager should be part of the host software package. The startup of BTChatd illustrated in rows 6–10 is also similar to that in Fig. 4.2. The Bluetooth devices used in the chat session have never met before, so they do not have a link key, as rows 11–12, 16–17, 21–22, and 26–27 illustrate. Because the piconet master is in the service-level enforced security mode (Bluetooth security mode 3 can also be used if desired), all slaves must enter the PIN code (see rows 12, 17, 22, and 27 in which the master requests each slave to enter the PIN code) that matches the master's PIN code. After this, the initialization key and the combination key can be generated (see Chap. 2). The master and each slave perform two-way authentication using the generated combination key and the results of authentication are used to generate the 128-bit encryption key. Then both ends of the link store the combination key (see rows 13–14, 18–19, 23–24, and 28–29) for later use (i.e., the devices are bonded) and encrypt all data transferred via air. The next time the same bonded devices communicate with the same master device, they

```
(1)  May 27 12:11:23 itmw-ope24 hcid[7374]: HCI daemon ver 2.4 started
(2)  May 27 12:11:23 itmw-ope24 hcid[7374]: Starting security manager 0
(3)  May 27 12:11:23 itmw-ope24 bluetooth: hcid startup succeeded
(4)  May 27 12:11:23 itmw-ope24 sdpd[7380]: sdpd v1.5 started
(5)  May 27 12:11:23 itmw-ope24 bluetooth: sdpd startup succeeded
(6)  May 27 12:11:25 itmw-ope24 btchatd: Serial Port service registered
(7)  May 27 12:11:25 itmw-ope24 btchatd: Logging chat to /tmp/btlog.txt
(8)  May 27 12:11:25 itmw-ope24 btchatd: BtCHATD, BlueZ RFCOMM chat server running!
(9)  May 27 12:11:25 itmw-ope24 btchatd: Waiting for connection on RFCOMM channel 10
(10) May 27 12:11:25 itmw-ope24 btchatd: Tuomas Kepanen, kepanen@cs.uku.fi
(11) May 27 12:11:40 itmw-ope24 hcid[7374]: link_key_request (sba=00:05:16:48:0C:FD, dba=00:05:4E:00:68:64)
(12) May 27 12:11:40 itmw-ope24 hcid[7374]: pin_code_request (sba=00:05:16:48:0C:FD, dba=00:05:4E:00:68:64)
(13) May 27 12:11:40 itmw-ope24 hcid[7374]: link_key_notify (sba=00:05:16:48:0C:FD)
(14) May 27 12:11:40 itmw-ope24 hcid[7374]: Saving link key 00:05:16:48:0C:FD 00:05:4E:00:68:64
(15) May 27 12:11:40 itmw-ope24 btchatd: Connection from [00:05:4E:00:68:64]
(16) May 27 12:14:14 itmw-ope24 hcid[7374]: link_key_request (sba=00:05:16:48:0C:FD, dba=00:05:16:48:0C:F5)
(17) May 27 12:14:14 itmw-ope24 hcid[7374]: pin_code_request (sba=00:05:16:48:0C:FD, dba=00:05:16:48:0C:F5)
(18) May 27 12:14:14 itmw-ope24 hcid[7374]: link_key_notify (sba=00:05:16:48:0C:FD)
(19) May 27 12:14:14 itmw-ope24 hcid[7374]: Saving link key 00:05:16:48:0C:FD 00:05:16:48:0C:F5
(20) May 27 12:14:14 itmw-ope24 btchatd: Connection from [00:05:16:48:0C:F5]
(21) May 27 12:16:13 itmw-ope24 hcid[7374]: link_key_request (sba=00:05:16:48:0C:FD, dba=00:05:16:48:12:4C)
(22) May 27 12:16:13 itmw-ope24 hcid[7374]: pin_code_request (sba=00:05:16:48:0C:FD, dba=00:05:16:48:12:4C)
(23) May 27 12:16:13 itmw-ope24 hcid[7374]: link_key_notify (sba=00:05:16:48:0C:FD)
(24) May 27 12:16:13 itmw-ope24 hcid[7374]: Saving link key 00:05:16:48:0C:FD 00:05:16:48:12:4C
(25) May 27 12:16:13 itmw-ope24 btchatd: Connection from [00:05:16:48:12:4C]
(26) May 27 12:17:02 itmw-ope24 hcid[7374]: link_key_request (sba=00:05:16:48:0C:FD, dba=00:05:16:48:0C:F2)
(27) May 27 12:17:02 itmw-ope24 hcid[7374]: pin_code_request (sba=00:05:16:48:0C:FD, dba=00:05:16:48:0C:F2)
(28) May 27 12:17:02 itmw-ope24 hcid[7374]: link_key_notify (sba=00:05:16:48:0C:FD)
(29) May 27 12:17:02 itmw-ope24 hcid[7374]: Saving link key 00:05:16:48:0C:FD 00:05:16:48:0C:F2
(30) May 27 12:17:02 itmw-ope24 btchatd: Connection from [00:05:16:48:0C:F2]
```

Fig. 4.5 The startup of the BlueZ protocol stack when encryption is used, the startup of BTChatd, and the connection establishments of the piconet slaves [2, 53]

will use the stored combination key for authentication and encryption key generation [2, 53].

Now the piconet slaves have a similar chat session to that illustrated in Fig. 4.3, but now the data is encrypted with 128-bit encryption keys. This makes the eavesdropper's work very hard as Table 3.3 in Sect. 3.2 illustrates. The eavesdropper has to synchronize again with the piconet master in order to intercept the packets exchanged via air. Even if encryption is used, all packets can be intercepted and stored for later cryptographical analysis [2, 53].

Figure 4.6 illustrates the results of an Interception of Packets attack when encryption is used. Now an eavesdropper cannot understand the contents of the intercepted data. The 16-bit CRC field calculated from the Baseband packet payload does not match the received CRC field, so our Bluetooth protocol analyzer displays the CRC field in red to indicate the mismatch between the CRC checksums. Now the eavesdropper knows for sure that the messages are encrypted [2, 53].

4.2 Integrity Threats

Reflection attacks [15] (also referred to as *Relay attacks*) are based on the impersonation of target devices. An attacker does not have to know any secret information, because she only relays (reflects) the received information from one target device to another during the authentication (see Fig. 2.2 in Chap. 2), i.e., a Reflection attack in Bluetooth can be seen as a type of MITM attack (see Chap. 2 and Sect. 5.2) against authentication, but not against encryption. The only information needed is

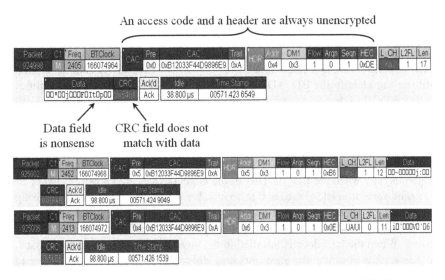

Fig. 4.6 The results of an Interception of Packets attack when encryption is used [2, 53]

the BD_ADDRs of the target devices. Reflection attacks are possible only when the target devices do not hear each other, i.e., communication between the target devices is not possible because they are out of each other's range, and the attacker has Bluetooth devices with adjustable BD_ADDRs (for example, protocol analyzers). Moreover, the attacker must be capable of relaying the received information between the devices that undergo the Reflection attack, because the target devices can be very far away from each other. These kinds of special conditions are not rare in Bluetooth networks, because Bluetooth is a short-range communication technology and it is very likely that devices occasionally move out of each other's range [2, 15].

There are two different kinds of Reflection attack: *One-Sided Reflection attack* and *Two-Sided Reflection attack*. Only one target device is impersonated in a One-Sided Reflection attack, while a Two-Sided Reflection attack requires that both target devices are impersonated. These two attacks and possible countermeasures are described in [15]. Bluetooth specifications up to 2.0+EDR do not provide any proper countermeasures against Reflection attacks, but Bluetooth versions 2.1+EDR, 3.0+HS, and 4.0 provide protection against these kinds of active eavesdropping attacks (MITM attacks). However, it has been shown that MITM attacks against Bluetooth 2.1+EDR/3.0+HS/4.0 devices are also possible [2, 19–23]. An attacker can successfully perform authentication using a Reflection attack, but she cannot continue the attack if the target devices want to have encrypted communication, i.e., the attacker does not know the secret PIN code, link key, or encryption key. Therefore, Reflection attacks are dangerous only when encryption is not used [2, 15].

A very dangerous form of integrity threat takes place when an attacker uses a stronger RF signal in order to displace the active piconet device. The main principle for successfully completing an *Exploitation of a stronger RF signal attack* [12] is to send the target device's receiver an RF signal that is at least 11 dB stronger than the signal that the legitimate piconet device is sending. The attack can be performed, for example, in the following way. Let us assume that the attacker A' wants to have a sensitive file F that is located on a server S. First, A' eavesdrops communication until she can identify the BD_ADDR of S on which F is located. A' also identifies the BD_ADDR of device A, which is authorized to access F. Second, A' waits until A connects to S and is properly authenticated. Third, A' captures the channel of A by impersonating A and transmitting signals that are at least 11 dB stronger at the receiver of S than those A was sending. Fourth, A' continues to pose as A, and finally A' requests the desired F from S. An exploitation of a stronger RF signal attack is dangerous only when the BD_ADDRs of the target devices are known [2, 12].

A *Backdoor attack* [34] means that an attacker establishes a trusted relationship with the target device through authentication and ensures that this trusted relationship no longer appears to be in the target device's register of paired (authenticated) devices. When the backdoor is installed in the target device via a Backdoor attack, the attacker can continue the attack in many different ways: for example, trying to exploit the resources of the target device via a trusted relationship, trying to perform a BlueSnarfing attack (see Sect. 4.1), or trying to slip a virus or worm to the target device (see Sect. 4.4). A Backdoor attack works only if the BD_ADDR of the target device is known. Moreover, the target device has to be vulnerable to a Backdoor attack. A list of the devices known to be vulnerable to Backdoor attacks without a firmware/software update can be found in [2, 34].

4.3 DoS Threats

DoS threats can be roughly divided into two parts: *attacks against the PHY* and *attacks against protocols above the PHY* [2, 34].

At the PHY, an attacker can jam the piconet entirely or capture the channel from the legitimate piconet device. Let us assume that a jammer wants to disrupt (jam) the PHY by hopping along with the piconet devices and sending random data in every timeslot. How far away can the jammer be? We can calculate the distance by making a number of realistic assumptions such as [2, 12]:

1. The target piconet devices are using omnidirectional antennas with 0 dBi gain.
2. The Carrier-to-Interference ratio (C/I ratio; also referred to as CIR) of the Bluetooth radio must be at least 0 dB for effective jamming, i.e., jamming power is at least equal to the desired signal's power at the target receiver.
3. All target piconet receivers have the desired signal power level of either −60 dBm or −40 dBm.
4. The jammer's transmit power is either 20 dBm (class 1 device) or 30 dBm (Bluetooth transmitter with a power amplifier).
5. The level of obstacles is either light (n = 2.5) or moderate (n = 3.0).
6. The jammer uses a directional antenna with a gain of 20 dBi.

Table 4.1 The range of susceptibility for a jammer on a target piconet [2, 12]

Level of obstacles:	n:	Jammer's TX power (dBm):	Minimum jamming signal power at target receiver (dBm):	Range of susceptibility for a jammer (m):
Light	2.5	20	−40	6
Light	2.5	30	−40	16
Light	2.5	20	−60	40
Light	2.5	30	−60	100
Moderate	3.0	20	−40	5
Moderate	3.0	30	−40	10
Moderate	3.0	20	−60	22
Moderate	3.0	30	−60	46

Table 4.1 presents the *range of susceptibility* for the jammer on a target piconet with these prerequisites calculated using the formula $d_j = 10^{(P_t - P_r - 40)/(10n)}$, where d_j is the range of susceptibility for the jammer, P_t is the jammer's transmit power, P_r is the minimum power needed at a target piconet receiver for jamming to occur, and n is the PL exponent. A more precise definition of the range of susceptibility calculations can be found in [2, 12].

As Table 4.1 illustrates, a jammer can perform a *Disruption of the PHY attack* [12] relatively far away from the communicating devices (up to 100 m) by using a Bluetooth transmitter with a power amplifier and a directional antenna with a gain of 20 dBi. This type of attack can be very dangerous if the attacker is using a stronger RF signal to displace the existing legitimate piconet device (see Sect. 4.2) and then trying to steal some sensitive information from the target device [2, 12].

Attacks on higher levels of the Bluetooth protocol stack try to exploit some of the characteristics of higher level protocols in an attempt to occupy the attention of one or more devices of the piconet in such a way that they are unable to serve other legitimate devices within a reasonable time. These kinds of attacks are not normally very dangerous, because an attacker does not steal any information from the target device. However, these kinds of attacks can be very annoying if the attacker uses them non-stop to deny the legitimate piconet devices access to the piconet services, or at least in such a way that they have considerably slowed throughput. Moreover, the attacker can use these kinds of attacks to mislead the target devices in such a way that they delete previously stored link keys so that the initial pairing process is restarted. For example, BD_ADDR Duplication attacks, SCO/eSCO attacks, Big Negative Acknowledgement (NAK) attacks, L2CAP Guaranteed Service attacks, Bluetooth OBEX Message attacks, BlueSmacking attacks, BlueSpamming attacks, and Battery Exhaustion attacks can be classified as attacks against protocols above the PHY [2].

A *BD_ADDR Duplication attack* [2, 12] is based on the idea that an attacker places a bug in the range of susceptibility. The bug duplicates the BD_ADDR of the target device. When any Bluetooth device tries to make a connection with the target device, either the target device or both devices, i.e., the target device and the bug, will respond and jam each other. In this way, the attacker has denied access from the legitimate device. The most effective way to perform this attack is to duplicate the BD_ADDR of the piconet master, because all information within the piconet goes through the master device. When the BD_ADDR of the piconet master is duplicated by the bug, all connections within that piconet will be effectively jammed due to simultaneous responses of both the target device and the bug [2, 12].

A BD_ADDR Duplication attack [2, 12] is normally possible only if the target device is configured as discoverable (see Chap. 2). However, as we explained earlier, there are ways to find non-discoverable devices. A BD_ADDR Duplication attack also requires that an attacker has a Bluetooth device with an adjustable BD_ADDR, because the bug must be capable of duplicating the BD_ADDR of the target device. Some commercially available Bluetooth protocol analyzers, such as LeCroy BTTracer/Trainer [42], support the BD_ADDR duplication feature. Therefore, it is not a problem for the attacker. Moreover, a BD_ADDR Duplication attack requires that the bug must be capable of impersonating the piconet master in order to respond to connection attempts of legitimate Bluetooth devices [2].

In our practical experiments [2, 55] we used an unmodified Bluetooth 1.1-compatible Nokia 6310i mobile phone [43] as the piconet master, a laptop connected to the LeCroy BTTracer/Trainer protocol analyzer [42] with one Bluetooth 1.1 compatible radio unit as the bug, and three Bluetooth headsets (Nokia's Bluetooth 1.1 compatible HDW-2 [56], Nokia's Bluetooth 2.0+EDR compatible HS-26W [57], and Sony Ericsson's Bluetooth 2.0+EDR compatible HBH-610 [58]) and Epox's Bluetooth 2.0+EDR compatible USB dongle BT-DG07A+ [59] as the legitimate piconet devices. LeCroy BTTracer/Trainer v2.2 software [42], which provides CATC Scripting Language [44] was also used [2, 55].

CATC Scripting Language was used to implement our Bluetooth security analysis tool, *BD_ADDR Duplication Security Analysis Tool* [2, 55], which was successfully used to perform BD_ADDR Duplication attacks. Our Bluetooth security analysis tool works in the following way [2, 55]:

1. We discover the BD_ADDR of the non-discoverable target mobile phone (the piconet master) by using a LeCroy BTTracer/Trainer protocol analyzer (the bug).
2. We use the bug to duplicate the BD_ADDR of the piconet master.
3. We use the bug to impersonate the piconet master. When any legitimate piconet device (the headsets or the USB dongle) tries to communicate with the piconet master, the bug responds each time simultaneously with the piconet master and therefore together they deny the legitimate piconet devices access by jamming each other.

Figure 4.7 (see rows 1–30) illustrates our practical experiment in which the bug successfully denies the legitimate piconet devices access. The bug first discovers the BD_ADDR value (see row 1) and the user-friendly name of the victim device (see rows 2–4). It impersonates the victim device by duplicating its BD_ADDR value (see rows 5–7), setting itself to require both authentication (see rows 8 and 13) and encryption (see row 14), and emulating the Headset Audio Gateway Profile that is

```
(1)   The BD_ADDR of the Piconet Master Is: 0002EEB0294D
(2)   HCI_Evt> Remote_Name_Request_Complete
(3)      BD_ADDR : 0002EEB0294D
(4)      Name      : "Nokia 6310i"
(5)   TCI_Evt> CATC_SetBdAddr_Complete
(6)      BD_ADDR              : 0002EEB0294D
(7)   The Bug Has Successfully Duplicated the BD_ADDR of the Piconet Master!
(8)   HCI_Evt> CATC_Write_PIN_Response_Enable_Complete
(9)   OBEX_Evt> ServerDeinit_Complete
(10)     Status                : Success
(11)  SDP_Evt> AddProfileServiceRecord_Complete
(12)     Profile               : Headset Audio Gateway
(13)  HCI_Evt> Write_Authentication_Enable_Complete
(14)  HCI_Evt> Write_Encryption_Mode_Complete
(15)  The Bug Has Successfully Impersonated the Piconet Master!
(16)  The Bug Is Waiting for a Connection..
(17)  HCI_Evt> Connection_Request
(18)     Incoming connection accepted
(19)  HCI_Evt> PIN_Code_Request
(20)     PIN reply             : 0000
(21)  HCI_Evt> Connection_Error
(22)     Incoming connection failed, reason : Authentication Failure
(23)  The Bug Is Waiting for a Connection..
(24)  HCI_Evt> Connection_Request
(25)     Incoming connection accepted
(26)  HCI_Evt> PIN_Code_Request
(27)     PIN reply             : 0000
(28)  HCI_Evt> Connection_Error
(29)     Incoming connection failed, reason : Authentication Failure
(30)  The Bug Is Waiting for a Connection..
```

Fig. 4.7 The result of our BD_ADDR Duplication attack [2, 55]

supported by the victim device (see rows 9–12 and 15). Finally, the bug waits for connections from the legitimate piconet devices (see rows 16, 23, and 30) and every time a connection request is received, the bug performs the authentication by using an incorrect PIN code, so all authentication attempts will fail (see rows 17–22 and 24–29) [2, 55].

A *SCO/eSCO attack* [2, 12] is based on the fact that a real-time two-way voice reserves a great deal of a Bluetooth piconet's attention so that the legitimate piconet devices do not receive the service within a reasonable time. The most effective way to perform this type of attack is to establish a SCO or an eSCO link with the piconet master, because all information within the piconet goes through the master device. A SCO/eSCO attack is possible when the target device has a fixed or short adjustable PIN code, its BD_ADDR is known to an attacker, and it has support for SCO or eSCO links (see Chap. 1). The secret PIN code of the target device can be discovered via an On-Line PIN Cracking attack (see Sect. 4.4) or Off-Line PIN Recovery attack (see Sect. 4.1). A SCO/eSCO attack is normally possible only if the target device is configured as discoverable (see Chap. 2). However, as we already discussed earlier, there are ways to find non-discoverable devices. Moreover, a SCO/eSCO attack requires that an attacker intercepts the traffic of the initial pairing process between two Bluetooth devices when they meet for the first time (see Chap. 2). As explained in Sect. 4.1, there are many different ways to arrange or force the target devices to repeat the initial pairing process, so this is not a big problem for the attacker [2, 12].

The most effective way to perform this attack is to establish a SCO link that uses *High-Quality Voice 1 (HV1)* packets, because in that way all the piconet master's attention is reserved for the attacker and the legitimate piconet devices do not receive service within a reasonable time [2, 55].

In our practical experiments we used an unmodified Bluetooth 1.1-compatible Nokia 6310i mobile phone [43] as the piconet master, a laptop connected to the LeCroy BTTracer/Trainer protocol analyzer [42] with one Bluetooth 1.1 compatible radio unit as the attacking device, and three Bluetooth headsets (Nokia's HDW-2 [56], Nokia's HS-26W [57], and SonyEricsson's HBH-610 [58]) as the legitimate piconet devices. LeCroy BTTracer/Trainer v2.2 software [42], which provides CATC Scripting Language [44] was also used [2, 55].

CATC Scripting Language was used to implement our Bluetooth security analysis tool, *SCO/eSCO Security Analysis Tool* [2, 55], which was successfully used to perform SCO attacks. Our Bluetooth security analysis tool works in the following way [2, 55]:

1. We discover the BD_ADDRs of the headsets (the legitimate piconet devices) and the target mobile phone (the piconet master) by using a LeCroy BTTracer/Trainer protocol analyzer (the attacking device).
2. We discover the fixed PIN codes of the headsets by using our On-Line PIN Cracking Security Analysis Tool (see Sect. 4.4).
3. We use the attacking device to intercept the traffic of the initial pairing process (see Chap. 2) between the piconet master and one legitimate piconet device (a headset).
4. We use the attacking device to duplicate the BD_ADDR of the legitimate piconet device.
5. We use the attacking device to authenticate itself with the piconet master by using the traffic of the initial pairing process that was intercepted in step 3.
6. We use the attacking device to open a two-way realtime HV1 SCO link with the piconet master. In this way, the legitimate piconet devices do not receive service within a reasonable time.

The same experiment was successfully performed with each of our Bluetooth headsets, i.e., three practical experiments were successfully performed with our SCO/eSCO Security Analysis Tool [2, 55].

A *Big NAK attack* [2, 12] is based on the idea of putting the target device on an endless retransmission loop so that the legitimate piconet devices have considerably slowed throughput. In this attack, an attacker requests any information from the target device and every time the requested information is received, the attacker sends NAK, i.e., the transmission has failed. The most effective way to perform this attack is to request information from the piconet master, because all information within the piconet goes through the master device. A Big NAK attack is possible if the target device is configured to respond to every information request, or if the attacking device is capable of impersonating one legitimate piconet device. In addition, the BD_ADDR of the target device (typically a piconet master) must be known to the attacker. As we already discussed earlier, there are many ways to find non-discoverable devices. Moreover, if the attack requires the impersonation of a legitimate piconet device, the secret PIN code that is used between the piconet master and the legitimate piconet device must be known to the attacker, and the attacker must also intercept the traffic of the initial pairing process between these two target devices when they meet for the first time (see Chap. 2). The PIN code of the target device can be discovered via an On-Line PIN Cracking attack (see Sect. 4.4) or an Off-Line PIN Recovery attack (see Sect. 4.1). As we already explained earlier, there are also many ways to arrange or force the target devices to repeat the initial pairing process [2, 12].

In our practical experiments we used an unmodified Bluetooth 1.1 compatible Nokia 6310i mobile phone [43] as the piconet master, a laptop connected to the LeCroy BTTracer/Trainer protocol analyzer [42] with one Bluetooth 1.1 compatible radio unit as the attacking device, and Epox's Bluetooth 2.0+EDR compatible USB dongle BTDG07A+ [59] as the legitimate piconet device. LeCroy BTTracer/Trainer v2.2 software [42], which provides CATC Scripting Language [44] was also used [2, 55].

We implemented our Bluetooth security analysis tool, *Big NAK Security Analysis Tool* [2, 55], using CATC Scripting Language [44]. The tool was successfully used to perform Big NAK attacks. Our security analysis tool works in the following way [2, 55]:

1. We discover the BD_ADDRs of the Bluetooth USB dongle (the legitimate piconet device) and the target mobile phone (the piconet master) by using a LeCroy BTTracer/Trainer protocol analyzer (the attacking device).
2. We discover the PIN code used between the piconet master and the legitimate piconet device by using our On-Line PIN Cracking Security Analysis Tool (see Sect. 4.4).
3. We use the attacking device to intercept the traffic of the initial pairing process between the piconet master and the legitimate piconet device.
4. We use the attacking device to duplicate the BD_ADDR of the legitimate piconet device.
5. We use the attacking device to authenticate itself with the piconet master by using the traffic of the initial pairing process that was intercepted in step 3.
6. We use the attacking device to request information from the piconet master and every time the requested information is received, the attacking device sends NAK. In this way, the attacking device puts the piconet master into an endless retransmission loop and thus the legitimate piconet devices do not receive service within a reasonable time or at least they have considerably slowed throughput.

An *L2CAP Guaranteed Service attack* [2, 12] is based on the idea that an attacker requests the highest possible data rate or the smallest possible latency from the target device so that all other connections are refused and all throughput is reserved for the attacker. An L2CAP Guaranteed Service attack is possible when the BD_ADDR of the target device is known. However, the success of the attack is implementation-dependent, because not all Bluetooth devices necessarily support the abovementioned optional L2CAP Quality-of-Service (QoS) features [2, 12].

A *Bluetooth OBEX Message attack* [60] is based on the Nokia 6310i [43] Bluetooth mobile phone's flaw (some other Bluetooth mobile phones may also be vulnerable) that allows an attacker to perform a remote DoS attack. This attack can be performed by sending invalid Bluetooth OBEX messages to the target device. As a result of this attack, the target device will lose its availability and may crash/reboot. A Bluetooth OBEX Message attack is possible when the BD_ADDR of the target device is known. Moreover, the target device has to be vulnerable to a Bluetooth OBEX Message attack [60, 61].

A *BlueSmacking attack* [62] is based on using the standard tools that are shipped with the BlueZ protocol stack [37]. An attacker can use a BlueSmacking attack to knock out some Bluetooth devices immediately. A BlueSmacking attack is possible when the BD_ADDR of the target device is known. Moreover, the target device has to be vulnerable to a BlueSmacking attack: for example, many of HP's iPAQ Personal Digital Assistants (PDAs) are vulnerable to a BlueSmacking attack [61, 62].

A *BlueSpamming attack* [63] is based on the idea that an attacker spams Bluetooth devices with arbitrary files if they support OBEX: *BlueSpam* [63] is a Palm OS application for performing BlueSpamming attacks. A BlueSpamming attack is possible when the target device supports OBEX and its BD_ADDR is known to the attacker [61, 63].

A *Battery Exhaustion attack* [2, 12] is based on the idea of occupying the target device in such a way that it rather quickly consumes the battery of the target device [2, 12].

4.4 Multithreats

There are also many attacks which cannot be classified as only one single threat. For example, BlueBugging attacks, Blooovering attacks, HeloMoto attacks, On-Line PIN Cracking attacks, BTKeylogging attacks, and BTVoiceBugging attacks can be classified as *disclosure and integrity threats*.

A *BlueBugging attack* [2, 64] means that an attacker connects to the target device (typically a Bluetooth mobile phone) without alerting its owner, steals some sensitive information, such as an entire phonebook, calendar notes, or text messages, and has full access to the GSM (Global System for Mobile communications) AT command set, i.e., a BlueBugging attack is based on the exploitation of AT commands. This means that the attacker can, in addition to stealing information, send text messages to premium numbers, initiate phone calls to premium numbers, write phonebook entries, connect to the Internet, set call forwards, try to slip a Bluetooth virus or worm to the target device, and many other things. A BlueBugging attack is even more dangerous than a BlueSnarfing attack (see Sect. 4.1), because the attacker can do almost anything with the vulnerable target device. A BlueBugging attack is possible and dangerous if the BD_ADDR of the target device is known to the attacker. Moreover, the Bluetooth protocol stack of the target device has to be poorly implemented, i.e., there are serious flaws in the authentication and data transfer mechanisms of some Bluetooth devices.

A list of the devices known to be vulnerable to a BlueBugging attack without a firmware/software update can be found in [34]. Several public BlueBugging tools exist: for example, *btxml* [65] and *Blooover* [2, 64, 66, 67].

Our practical experiment, *BlueBugging attack* [2, 61], demonstrates the dangerousness of such an attack. The equipment needed for the practical experiment was a laptop with Linux Fedora Core 3 and BlueZ protocol stack [37] installed, one Bluetooth 1.1 compatible USB dongle, and an unmodified Bluetooth 1.1 compatible Nokia 6310i mobile phone [43]. In addition, a special tool, btxml [65], was installed on the laptop. Andreas Oberritter's btxml is a tool for a BlueBugging attack. It is capable of stealing the contents of the target mobile phone via Bluetooth and outputting the data in a standard Extensible Markup Language (XML) format. Originally, btxml was designed to work for Nokia 6310 and 6310i mobile phones, but it also works for Ericsson T610 and T68i mobile phones, and may work for some other Bluetooth mobile phones as well. It simply uses GSM AT commands over an RFCOMM connection, and no initial pairing process is required between the attacking device and the target device [2, 61].

The purpose of this practical experiment was to demonstrate the dangerousness of a BlueBugging attack and to determine the average time required for the attack by using btxml [65]. Figure 4.8 illustrates the results of this experiment. BD_ADDR (see row 2), a user-friendly name (a 1–248 byte user-defined string describing a Bluetooth device; see row 2), the device manufacturer (see row 3), device model (see row 4), firmware version (see row 5), International Mobile Equipment Identity (IMEI) code (see row 6), the entire phonebook (see rows 7–20), and all text messages (see rows 21–26) were easily discovered and stolen. IMEI can be used for illegal mobile phone cloning [2, 61].

A BlueBugging attack was repeated 50 times and the average time required for one successful attack when the target mobile phone had three contact numbers and two text messages stored was about 10.7 s. For half of the time btxml was performing an inquiry scan, i.e., searching for Bluetooth devices in range. Therefore, the actual time for connecting and stealing information is only about 5 s per BlueBugging attack if the BD_ADDR of the target device is known beforehand. Based on the results of this practical experiment, we can assume that the average time required for a BlueBugging attack varies from five seconds up to several minutes, depending on the amount of information, such as pictures, music files, text messages, phonebook entries, and calendar notes, stored on the vulnerable Bluetooth device. A BlueBugging attack is very dangerous, because millions of vulnerable Bluetooth devices [33, 34, 45–48], especially Bluetooth mobile phones, are used every day all over the world [2, 61].

Blooover [66] and its successor Blooover II [67] are derived from Bluetooth Hoover, because they run on handheld devices, such as PDAs or mobile phones, and are capable of stealing sensitive information by using a BlueBugging attack [2, 64]. A *Bloovering attack* [66, 67] can be done secretly by using only a Bluetooth mobile phone with Blooover or Blooover II installed. Blooover and Blooover II run on almost every Java 2 Micro Edition (J2ME) compatible handheld device. They are intended to serve as auditing tools, which can be used for checking whether Bluetooth devices are vulnerable or not, but they can be used for attacks against Bluetooth devices as

```
(1)  <?xml version="1.0" encoding="UTF-8"?>
(2)  <phone btaddr="00:02:EE:B0:29:4D" name="Nokia 6310i">
(3)          <manufacturer>Nokia</manufacturer>
(4)          <model>Nokia 6310i</model>
(5)          <revision>V5.50   03-03-03 NPL-1 (c) NMP. </revision>
(6)          <imei>351453208359469</imei>
(7)          <phonebook name="ME" size="500">
(8)                  <contact>
(9)                          <name>Test contact number</name>
(10)                         <number>+358501234567</number>
(11)                 </contact>
(12)                 <contact>
(13)                         <name>Another contact number</name>
(14)                         <number>+358447654321</number>
(15)                 </contact>
(16)                 <contact>
(17)                         <name>Yet another contact number</name>
(18)                         <number>+358112233445</number>
(19)                 </contact>
(20)         </phonebook>
(21)         <msgstorage name="ME">
(22)                 <message>"STO UNSENT","", This is a test message for btxml program..</message>
(23)                 <message>"STO UNSENT","", This is another test message...</message>
(24)         </msgstorage>
(25)         <msgstorage name="SM">
(26)         </msgstorage>
(27) </phone>
```

Fig. 4.8 The results of a BlueBugging attack using btxml [2, 61]

well. The source codes of Blooover and Blooover II have not been published (only
binaries). A Blooovering attack is dangerous only if the target device is vulnerable
to BlueBugging. Moreover, the attacker has to know the BD_ADDR of the target
device [2, 66, 67].

A *HeloMoto attack* [68] is a combination of BlueSnarfing (see Sect. 4.1) and Blue-
Bugging attacks. It was first discovered by Adam Laurie on Motorola's Bluetooth
mobile phones. A HeloMoto attack is based on the poorly implemented handling
of the trusted devices on some of Motorola's Bluetooth mobile phones (for example, models V80, V5xx, V6xx, and E398 are vulnerable without firmware/ software
update), and it gives full access to the GSM AT command set. A HeloMoto attack
is possible when the BD_ADDR of the target device is known. Moreover, the target
device has to be vulnerable to HeloMoto attack [61, 68].

An *On-Line PIN Cracking attack* [2, 35] means that an attacker is trying to connect
with the target device by guessing different PIN values. It is based on the idea of
changing the BD_ADDR of the attacking device every time a PIN guess fails, i.e.,
the attacker bypasses the ever-increasing delay between retries. An On-Line PIN
Cracking attack works only when the target device has a fixed or short adjustable PIN
code and its BD_ADDR is known to the attacker. Bluetooth versions up to 2.0+EDR
do not provide any proper countermeasures against On-Line PIN Cracking attacks,
while Bluetooth versions 2.1+EDR, 3.0+HS, and 4.0 provide SSP (see Chap. 2) to
protect against such attacks [2, 35].

Our On-Line PIN Cracking Security Analysis Tools, *On-Line PIN Cracking Script*
[2, 41] and *On-Line PIN Cracking Tool* [2, 55], are (as far as we know) the only security analysis tools for an On-Line PIN Cracking attack implemented so far [2, 41, 55].

In our first experiment, we successfully performed an On-Line PIN Cracking
attack by using a laptop connected to LeCroy's Bluetooth protocol analyzer [42]
with one Bluetooth 1.1 compatible radio unit and Nokia's Bluetooth 1.1 compatible

Wireless Headset HDW-2 [56]. A second radio unit will not speed up the process, because only one PIN trial can be performed with the same headset at the same time. However, it can be used for On-Line PIN Cracking with another headset or another Bluetooth device that has a fixed PIN code. LeCroy BTTracer/Trainer v2.2 software [42], which provides CATC Scripting Language [44] was also used in this practical experiment [2, 41].

CATC Scripting Language was used for creating our *On-Line PIN Cracking Script*, which works in the following way [2, 41]:

1. Change the local BD_ADDR of the protocol analyzer and set a PIN value for the next PIN trial. In this way, the ever increasing delay between retries is bypassed by changing the BD_ADDR of the attacking device every time a PIN guess fails.
2. Create a basic ACL link between the protocol analyzer and the target device.
3. Perform authentication with the target device by using the PIN value set in step 1. If authentication fails, go back to step 1. Otherwise, On-Line PIN Cracking has been completed successfully.

The success of our practical experiment is based on the fact that most Bluetooth headsets use only four-digit fixed PIN codes. Figure 4.9 illustrates an example of a successful On-Line PIN Cracking attack using our On-Line PIN Cracking Script. The protocol analyzer is set to require authentication for each connection with the target device (see row 1). The BD_ADDR value of the protocol analyzer is changed to a new value after every failed authentication attempt (see rows 2–3, 8–9, and 14–15). Two failed authentication attempts are performed with the target device (see rows 4–7 and 10–13). The third authentication attempt is successful (see rows 16–22) and therefore disconnection with the target device can be performed (see rows 23–25). Now the attacker has discovered the fixed PIN code of the target device and further attacks (see Sects. 4.1–4.4) against that device can be performed [2, 41].

In our practical experiment, the average time required for one PIN trial was 4.7 s including the BD_ADDR change after every PIN trial. Therefore, the average On-Line PIN Cracking time, i.e., the time for 5,000 PIN trials using Bluetooth 1.1 compatible devices was 6.5 h ($5,000 \times 4.7\,s = 23,500\,s \approx 6.5\,h$). The same practical experiment with Bluetooth 1.2 or 2.0+EDR devices would have been faster to perform, because Bluetooth specifications 1.2 and 2.0+EDR support faster (less than two seconds per device) connection establishment. Assuming that the average time required for one PIN trial using Bluetooth 1.2 or 2.0+EDR compatible devices is 2.0 s, the average On-Line PIN Cracking time for all 5000 PIN trials can be as short as 2.8 hours ($5000 \times 2.0\,s = 10000\,s \approx 2.8\,h$) [2, 41].

Since our On-Line PIN Cracking Script runs only in Windows environments and requires expensive special hardware, we decided to implement another version of the On-Line PIN Cracking Security Analysis Tool in order to eliminate these restrictions: our *On-Line PIN Cracking Tool* [2, 55] works in Linux environments. It requires the BlueZ protocol stack [37] and at least one Bluetooth USB dongle to work, i.e., an expensive Bluetooth protocol analyzer is not required. The best performance with our On-Line PIN Cracking Tool can be achieved when two Bluetooth USB dongles are used simultaneously for an On-Line PIN Cracking attack, i.e., the second

```
(1)  HCI_Evt> Write_Authentication_Enable_Complete
(2)  TCI_Evt> CATC_SetBdAddr_Complete
(3)     BD_ADDR              : 000000002330
(4)  HCI_Evt> PIN_Code_Request
(5)     PIN reply            : 2330
(6)  HCI_Evt> Connection_Error
(7)     Error                : Authentication Failure
(8)  TCI_Evt> CATC_SetBdAddr_Complete
(9)     BD_ADDR              : 000000002331
(10) HCI_Evt> PIN_Code_Request
(11)    PIN reply            : 2331
(12) HCI_Evt> Connection_Error
(13)    Error                : Authentication Failure
(14) TCI_Evt> CATC_SetBdAddr_Complete
(15)    BD_ADDR              : 000000002332
(16) HCI_Evt> PIN_Code_Request
(17)    PIN reply            : 2332
(18) HCI_Evt> Pairing_Complete
(19)    BD_ADDR              : 00038935446F
(20) HCI_Evt> Connection_Complete
(21)    BD_ADDR              : 00038935446F
(22)    HCI Handle           : 0x000B
(23) HCI_Evt> Disconnection_Complete
(24)    BD_ADDR              : 00038935446F
(25)    Reason               : No Connection
```

Fig. 4.9 An example of a successful On-Line PIN Cracking attack [2, 41]

USB dongle changes its BD_ADDR while the first USB dongle is performing the On-Line PIN Cracking attack and vice versa. Our On-Line PIN Cracking Tool uses the standard tools shipped with BlueZ and our own modifications of some BlueZ tools. Manufacturer-specific commands are used to change the BD_ADDR of the attacking device every time the PIN guess fails, i.e., not all Bluetooth USB dongles are supported by our security analysis tool. The supported manufacturers are Cambridge Silicon Radio (CSR), Texas Instruments (TI), Ericsson, and Zeevo, i.e., most Bluetooth USB dongles on the market support our On-Line PIN Cracking Tool [2, 55].

Changing the BD_ADDR value of a typical Bluetooth USB dongle takes only an average of two seconds, which is much faster than the time required for one PIN trial. This allows us to use a simple parallelization technique to speed up the attack, which works as follows. The average time for one PIN trial (excluding the BD_ADDR change) with our On-Line PIN Cracking Tool is ten seconds for the "old" Bluetooth 1.0/1.1 devices and four seconds for the "new" Bluetooth 2.0+EDR devices, which support faster connection establishment. Therefore, a second Bluetooth USB dongle can be used to save an average of two seconds per PIN trial, i.e., the attack can be performed 17–33 % faster ($100\% \times 2/12 = 17\%$ and $100\% \times 2/6 = 33\%$) by using two Bluetooth USB dongles in parallel. Figure 4.10 illustrates our On-Line PIN Cracking Tool in action [2, 55].

As Fig. 4.10 illustrates, the On-Line PIN Cracking Tool works in a similar way to the On-Line PIN Cracking Script described in Fig. 4.9. The local BD_ADDR value of the attacking device is changed to a new value after every failed authentication

Fig. 4.10 The On-Line PIN
Cracking Tool in action [2, 55]

```
(1)   PIN Code for the Next PIN Trial Is: 1232
(2)   Connecting to the Remote Bluetooth Device..
(3)   Can't create connection: Input/output error
(4)   New (Local) BD_ADDR Will Be: 29:29:29:29:29:29
(5)   Local BD_ADDR Has Been Changed to New Value!
(6)   Device Reset Has Been Completed Successfully!
(7)   PIN Code for the Next PIN Trial Is: 1233
(8)   Connecting to the Remote Bluetooth Device..
(9)   Can't create connection: Input/output error
(10)  New (Local) BD_ADDR Will Be: 30:30:30:30:30:30
(11)  Local BD_ADDR Has Been Changed to New Value!
(12)  Device Reset Has Been Completed Successfully!
(13)  PIN Code for the Next PIN Trial Is: 1234
(14)  Connecting to the Remote Bluetooth Device..
(15)  Authentication Has Been Successfully Completed!
(16)  PIN Code of the Remote Bluetooth Device Is: 1234
```

attempt (see rows 4–6 and 10–12). Two failed authentication attempts are performed with the target device (see rows 1–3 and 7–9). The third authentication attempt is successful (see rows 13–16) and therefore the attacker has discovered the secret PIN code of the target device. The On-Line PIN Cracking Script is faster than the On-Line PIN Cracking Tool, because it runs on special hardware, LeCroy's Bluetooth protocol analyzer [42], which can use a Bluetooth radio much more efficiently than a normal PC with a Bluetooth USB dongle. On the other hand, it is also a much more expensive approach to On-Line PIN Cracking, thus making our On-Line PIN Cracking Tool a very economical solution [2, 55].

An On-Line PIN Cracking attack is very feasible and dangerous if the fixed PIN code of the target device is short and has no case-sensitive alphanumerical characters (and perhaps some other characters as well). In addition, an attacker does not necessarily need to go through all PIN values in one day. She can continue the attack some other day when the target device is back within the range of vulnerability. Since Bluetooth specifications up to 2.0+EDR do not provide any proper countermeasures for On-Line PIN Cracking attacks, all countermeasures (if any) are up to the device manufacturer [2, 55].

Our Bluetooth security attack, a *BTKeylogging attack* [2, 41], extends both the Brute-Force BD_ADDR Scanning attack [2, 35] and the On-Line PIN Cracking attack [2, 35]. A BTKeylogging attack is carried out on a wireless connection between a Bluetooth-enabled keyboard (which is also used for typing a PIN code during the initial pairing process) and a PC: in the attack, an attacker uses the target device (a Bluetooth-enabled keyboard) as a "Bluetooth keylogger" by intercepting all packets (i.e., all keystrokes) sent via air and decrypting them. The attack is possible when the target keyboard has a fixed or short adjustable PIN code and its BD_ADDR is known to the attacker. Moreover, the attacker must witness the initial pairing process between the target keyboard and the target computer: thereafter, all intercepted information can be decrypted. As we discussed earlier, there are different ways to arrange or force target devices to repeat the initial pairing process (see Sect. 4.1) [2, 41].

A BTKeylogging attack was performed in the following way. We discovered the BD_ADDRs of the target devices via a Brute-Force BD_ADDR Scanning attack and we also discovered the fixed PIN code of the target keyboard via an

On-Line PIN Cracking attack. We also used a Bluetooth protocol analyzer to intercept all the required information (the IN_RAND value, LK_RAND values, AU_RAND value, and EN_RAND value) for the BTKeylogging attack. Then the keyboard was used as a keylogger by intercepting all keypresses. We also successfully decrypted all the intercepted information. As described in Chap. 2, K_{init} can be produced using the formula $K_{init} = E_{22}(PIN',L',IN\_RAND)$. This can be used to decrypt the intercepted LK_RAND values (LK_RAND$_A$ and LK_RAND$_B$), i.e., (LK_RAND$\oplus K_{init}$)$\oplus K_{init}$ = LK_RAND. K_{AB} can be produced using the formula $K_{AB} = K_A \oplus K_B = E_{21}(BD\_ADDR_A, LK\_RAND_A) \oplus E_{21}(BD\_ADDR_B,$ LK_RAND$_B$), and ACO can be produced using the $E_1(AU\_RAND_A, BD\_ADDR_B,$ $K_{AB})$ function. K_C can be produced using the formula $K_C = E_3(EN\_RAND_A, ACO,$ $K_{AB})$, and finally the keystream can be generated using the $E_0(K_C, CLK_{26-1},$ BD_ADDR$_A$) function. Each intercepted Baseband packet can be decrypted by XORing its encrypted payload with the correct keystream, i.e., Ciphertext$\oplus$Keystream = (Plaintext$\oplus$Keystream)$\oplus$Keystream = Plaintext. It is also worth noting that each intercepted Baseband packet must be stamped with the associated CLK_{26-1} value before it is stored (most commercially available Bluetooth protocol analyzers automatically support this feature) in order to produce the correct keystream for each Baseband packet [2, 41].

Another way of performing a BTKeylogging attack is to use an Off-Line PIN Recovery attack (see Sect. 4.1) instead of an On-Line PIN Cracking attack to discover the fixed or short adjustable PIN code of the target keyboard. This approach requires that the attacker intercepts the IN_RAND value, LK_RAND values, AU_RAND value, SRES value, and EN_RAND value, i.e., it requires that the attacker intercepts the SRES value in addition to the values in the example described above. Then the attacker tries to calculate the correct SRES value by guessing different PIN values until the calculated SRES equals the intercepted SRES. The SRES value can be produced by the $E_1(AU\_RAND_A, BD\_ADDR_B, K_{AB})$ function, i.e., the same function that produces ACO. As described in Sect. 4.1, an SRES match does not necessarily guarantee that the attacker has discovered the correct PIN code, but the chances are quite high especially if the PIN code is short. The other phases of a BTKeylogging attack in this approach are the same as described in our example above [2, 41].

Some Bluetooth keyboards allow users to change the PIN code of the keyboard, but unfortunately in most cases this new secret PIN code has to be typed at the computer side. Then it is sent via the Bluetooth link to the keyboard and stored. This is clearly a big security risk, because an attacker can intercept the new PIN code by using a Bluetooth protocol analyzer. This means that the attacker will not even need to perform an On-Line PIN Cracking attack or Off-Line PIN Recovery attack to discover the new PIN code [2, 41].

Clearly, BTKeylogging attacks can be very dangerous, because an attacker can intercept and decrypt all keystrokes of the victim user: for example, all usernames and passwords as well as all sent e-mails can be easily stolen by using the victim keyboard as a "Bluetooth keylogger".

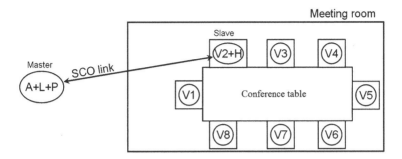

Fig. 4.11 The first BTVoiceBugging attack scenario [2, 41]

Our Bluetooth security attack, a *BTVoiceBugging attack* [2, 41], extends both a Brute-Force BD_ADDR Scanning attack and an On-Line PIN Cracking attack. A BTVoiceBugging attack is possible when the target device has a fixed or short adjustable PIN code, its BD_ADDR is known to the attacker, and it has support for SCO or eSCO links (see Chap. 1). In the attack, the attacker uses the target device (for example, a Bluetooth headset or a Bluetooth-enabled PC/laptop equipped with a microphone and speakers) as a "Bluetooth bugging device". In such a case, the attacker can listen to sensitive conversations (for example, important business or other meetings taking place in the vicinity of the target device) via a SCO or an eSCO link, and she can also record these conversations for later use. Because the link between the attacking device and the target device is two-way, it is also possible to send voice packets, such as a lubricious voice message or a funny song, to the target device. However, a BTVoiceBugging attack does not make it possible to eavesdrop on the voice communication of another SCO or eSCO link that the same target device is using simultaneously with another device. Therefore, in order to eavesdrop on the voice communication of two target devices, additional attacks, such as an Off-Line Encryption Key Recovery attack and Interception of Packets attack (see Sect. 4.1), are required [2, 41].

We performed a BTVoiceBugging attack in the laboratory environment in the following way. We discovered the BD_ADDR of the target headset via a Brute-Force BD_ADDR Scanning attack and we also discovered the fixed PIN code of the target headset via an On-Line PIN Cracking attack, i.e., we obtained all the information required for a BTVoiceBugging attack. Then we used a Bluetooth protocol analyzer to open a two-way real-time SCO connection with the target headset, i.e., the headset was used as a bugging device. We also intercepted all voice packets and exported them to a Waveform (WAV) file that was stored for later use [2, 41].

We defined three different BTVoiceBugging attack scenarios in order to eavesdrop on a typical business meeting (see Figs. 4.11, 4.12 and 4.13). The first scenario is illustrated in Fig. 4.11 [2, 41].

In this scenario, the attacker (A) has a laptop (L) and a Bluetooth protocol analyzer (P). Seven participants (victims V1, V3, V4, V5, V6, V7, and V8) in the business meeting do not have any Bluetooth devices with them. However, one participant (V2)

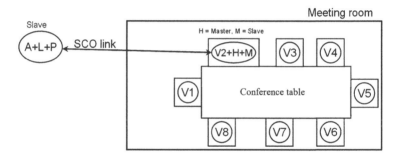

Fig. 4.12 The second BTVoiceBugging attack scenario [2, 41]

has in his pocket a Bluetooth headset (H) or any other Bluetooth device that supports a SCO/eSCO link and is also equipped with a microphone/speakers to enable a real-time two-way Bluetooth voice link, i.e., V2 is not aware that he has a "Bluetooth bugging device" in his pocket. The attacking device (L+P) must be a piconet master, because it must initiate the connection with V2's headset (H). Correspondingly, H is a piconet slave that establishes a real-time two-way SCO link with the piconet master (L+P) [2, 41].

The second BTVoiceBugging attack scenario is illustrated in Fig. 4.12. A, L, P, V1, V3, V4, V5, V6, V7, and V8 in this scenario are the same as in the first scenario. A meeting participant V2 has a Bluetooth headset (H) and a Bluetooth-enabled mobile phone (M) in his pocket. H is connected to M and therefore V2 is able to receive calls wirelessly via H. When V2 is not wirelessly receiving a call, there exists only a basic ACL link between H and M, i.e., a SCO link is automatically established only when a call is received. Therefore, if H supports multiple connections, A can establish a real-time two-way SCO link with it and thus V2 again has a "Bluetooth bugging device" in his pocket. H must be a piconet master, because it must automatically initiate a connection with M when the power is switched on. Correspondingly, M is a piconet slave that establishes a basic ACL link with the piconet master (H). The attacking device (L+P) is also a piconet slave, because it joins the existing piconet in which H is the piconet master [2, 41].

The third BTVoiceBugging attack scenario is illustrated in Fig. 4.13. A, L, and P in this scenario are the same as in the first scenario. None of the eight participants (victims V1, V2, V3, V4, V5, V6, V7, and V8) in the business meeting has any Bluetooth devices on her. However, there is a Bluetooth-enabled computer (C) equipped with a microphone (M) and speakers (S) to enable a real-time two-way Bluetooth voice link, i.e., the participants (victims) are not aware that a "Bluetooth bugging device" is on the table of the meeting room. C can be a PC in the meeting room, used to give PowerPoint presentations via a video projector or video-conferencing services via the Internet. Alternatively, C can be a Bluetooth-enabled laptop or PDA belonging to one of the participants (victims), equipped with M and S, which has been taken into the meeting room "just in case", i.e., it is not necessary that a participant uses C (or any Bluetooth services) during the meeting, but its power must be switched

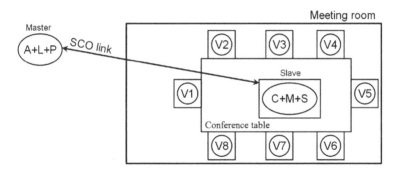

Fig. 4.13 The third BTVoiceBugging attack scenario [2, 41]

on. The attacking device (L+P) must be a piconet master, because it must initiate a connection with C. Correspondingly, C is a piconet slave that establishes a real-time two-way SCO link with the piconet master (L+P) [2, 41].

Clearly, BTVoiceBugging attacks can be very dangerous, because an attacker can listen to sensitive conversations taking place in the vicinity of the "Bluetooth bugging device" and he can also record these conversations for later use.

For example, BTPrinterBugging attacks and attacks based on using Bluetooth worms/viruses can be classified as *disclosure, integrity, and DoS threats*.

New Bluetooth applications create new security threats, for example, in printing. Bluetooth printers and printer adapters are widely used all over the world, especially in home environments. Sensitive information, such as documents containing personal identification or social security numbers, documents relating to business issues, and web pages relating to bank account information, are printed via Bluetooth without consideration of the possible security risks. Moreover, Bluetooth-enabled printers are in most cases always powered on, allowing long-lasting attacks. The typical communication range of a Bluetooth-enabled printer is up to 100 m indoors, because in most cases it is a class 1 device (see Sect. 3.1). Moreover, most Bluetooth-enabled printers have an enhanced sensitivity level, such as −80 dBm or better, and the communication range of such printers is even higher than that of printers with a standard sensitivity level. This means that attacks against Bluetooth-enabled printers can typically be carried out relatively far away from the targets. Most Bluetooth-enabled printers are configured as discoverable devices (see Chap. 2) by the fixed factory setting, i.e., it is not possible to configure the printer as non-discoverable. Therefore, the BD_ADDR of a Bluetooth-enabled printer can typically be discovered in a few seconds by an attacker [2].

Our *BTPrinterBugging attacks* [2, 69] are based on the idea that an attacker abuses the target Bluetooth-enabled printer in order to do various harmful things. The attacker can, for example, both intercept and decrypt all the information that is sent to the printer (a *BTPrinterBugging via Packet Interception attack* [2, 69]), use the printer remotely as if it was her own (a *BTPrinterBugging via Impersonation attack* [2, 69]), deny access to the printer to the legitimate piconet users (a *BTPrint-*

erBugging via Access Denial attack [2, 69]), and do many other harmful things. We have demonstrated experimentally that attacks against Bluetooth-enabled printers become practical by using our security analysis tools [2, 69]. The attacks make use of our efficient implementations of security analysis tools [2, 69].

If an attacker wants to intercept and decrypt all the information that is sent to a Bluetooth-enabled printer via air, the BD_ADDR of the printer, the BD_ADDR of the legitimate piconet device that is using the printer, and the secret PIN code that is used between these two target devices must be known to the attacker. Moreover, the attacker must intercept the traffic of the initial pairing process between these two devices when they meet for the first time (see Chap. 2). This kind of attack is normally possible only if the target devices are configured as discoverable devices (see Chap. 2). However, there are ways to find non-discoverable devices (see Sect. 4.1) [2, 69].

Most Bluetooth-enabled printers have only four-digit fixed or short adjustable PIN codes, and in many cases this PIN code is as naive as four zeros, i.e., "0000" in ASCII, which is definitely almost every attacker's first PIN code guess. In addition, based on the "user-friendly" name and the company_id value (see Chap. 2) of a Bluetooth-enabled printer, the attacker can determine its manufacturer and the model. With this information, it is very easy for the attacker to download the user manual of the printer from the Internet in order to find out the required fixed PIN code. Moreover, many Bluetooth-enabled printers have no support for a PIN code at all, i.e., no PIN code is available. According to the Bluetooth specification [1], the default value of zero, i.e., "0" in ASCII, will be used as a PIN code for Bluetooth security operations if no PIN code is available. Therefore, "0" is also a very good PIN guess for the attacker. Clearly then, the PIN code of a Bluetooth-enabled printer is in most cases very easy to guess. However, some Bluetooth-enabled printers may have fixed or adjustable PIN codes, which are very hard to guess. In such a case, an On-Line PIN Cracking attack or Off-Line PIN Recovery attack can be used to discover PIN codes. As described in Sect. 4.1, there are many ways to arrange or force target devices to repeat the initial pairing process. Therefore, the requirement to intercept the traffic of the initial pairing process between two target devices is not a big problem for an attacker [2, 69].

We designed, implemented, and tested a tool, the *BTPrinterBugging via Packet Interception Security Analysis Tool* [2, 69], which was successfully used to perform BTPrinterBugging via Packet Interception attacks against four Bluetooth 1.1 compatible USB printer adapter models: Conceptronic [70], Mentor [71], Tecom [72], and Belkin [73]. All these printer adapters are configured as discoverable devices by the fixed factory setting. In addition, Conceptronic and Tecom printer adapters have no support for a PIN code at all. Therefore, the default PIN code "0" is used when Bluetooth security operations are required with them. The PIN codes for Mentor and Belkin printer adapters are "1234" and "belkin", respectively. Each of these so-called "secret PIN codes" can easily be found in the printer adapter manual, which can be freely downloaded from the Internet. Because all four printer adapters have either default or fixed PIN codes, all Bluetooth-enabled devices that use them via an encrypted link must also use the same PIN code—otherwise printing is not pos-

```
Bookmark,Frame#,Role,Addr.,HCRP Data,Frame Size,Delta,Timestamp,
,1969,Master,2,0x 1b 25 2d 31 32 33 34 35 58 40 50 4a 4c 20 4a 4f 42 20 4e 41
4d 45 3d 22 48 69 67 68 6c 79 43 6c 61 73 73 69 66 69 65 64 44 6f 63 75 6d 65
6e 74 32 2e 70 64 66 22 0a 40 50 4a 4c 20 53 45 54 20 53 54 52 49 4e 47 43 4f
44 45 53 45 54 3d 55 54 46 38 0a 40 50 4a 4c 20 43 4f 4d 4d 45 4e 54 20 22 48
50 20 43 6f 6c 6f 72 20 4c 61 73 65 72 4a 65 74 20 34 36 35 30 20 50 43 4c 20
36 20 28 42,147, 00:00:02.845179,1.12.2006 10:15:03.249720 ,
,1970,Master,2,0x 65 6c 6b 69 6e 29 20 28 36 30 2e 33 32 2e 31 30 31 2e 34 31
29 3b 20 4d 69 63 72 6f 6f 66 74 20 57 69 6e 64 6f 77 73 20 58 50 20 35 2e
31 2e 32 36 30 30 2e 31 3b 20 55 6e 69 64 72 76 20 30 2e 33 2e 31 32 39 36 2e
31 22 0a 40 50 4a 4c 20 43 4f 4d 4d 45 4e 54 20 22 55 73 65 72 6e 61 6d 65 3a
20 57 4b 53 41 44 4d 49 4e 3b 20 41 70 70 20 46 69 6c 65 6e 61 6d 65 3a 20 48
69 67 68 6c,147, 00:00:00.005000,1.12.2006 10:15:03.254720 ,
,1971,Master,2,0x 79 43 6c 61 73 73 69 66 69 65 64 44 6f 63 75 6d 65 6e 74 32
2e 70 64 66 3b 20 31 32 2d 31 2d 32 30 30 36 22 0a 40 50 4a 4c 20 53 45 54 20
4a 4f 42 41 54 54 52 3d 22 4a 6f 62 41 63 63 74 31 3d 57 4b 53 41 44 4d 49 4e
22 0a 40 50 4a 4c 20 53 45 54 20 4a 4f 42 41 54 54 52 3d 22 4a 6f 62 41 63 63
74 32 3d 54 4b 54 2d 4c 54 2d 4d 35 31 33 32 22 0a 40 50 4a 4c 20 53 45 54 20
4a 4f 42 41,147, 00:00:00.005000,1.12.2006 10:15:03.259720 ,
,1972,Master,2,0x 54 54 52 3d 22 4a 6f 62 41 63 63 74 33 3d 54 4b 54 2d 4c 54
2d 4d 35 31 33 32 22 0a 40 50 4a 4c 20 53 45 54 20 4a 4f 42 41 54 54 52 3d 22
4a 6f 62 41 63 63 74 34 3d 32 30 30 36 31 32 30 31 31 30 31 34 35 32 22 0a 40
50 4a 4c 20 44 4d 49 4e 46 4f 20 41 53 43 49 49 48 45 58 3d 22 30 34 30 30 30
34 30 31 30 31 30 32 30 44 31 30 31 30 30 31 31 35 33 32 33 30 33 30 33 36 33
31 33 32 33 33,147, 00:00:00.006250,1.12.2006 10:15:03.265970 ,
```

Fig. 4.14 An example of a comma-separated ASCII text file [2, 69]

sible. Moreover, all four printer adapters are class 1 Bluetooth devices with enhanced sensitivity levels, making long-distance attacks against them possible [2, 69].

In our practical experiments we used a USB printer with a Bluetooth USB printer adapter as the Bluetooth-enabled printer, a laptop running a Frontline FTS4BT [74] protocol analyzer with one Bluetooth 2.0+EDR compatible USB ComProbe as the attacking device, and a Bluetooth-enabled laptop (Bluetooth 1.2 compatible) as the legitimate Bluetooth-enabled device that was using the printer. Frontline FTS4BT v6.10.4.0 software [74] was also used in our practical experiments. Our Bluetooth security analysis tool works in the following way [2, 69]:

1. All information that is sent to the Bluetooth-enabled printer is intercepted by using a Frontline FTS4BT protocol analyzer. If the intercepted data are encrypted, they are automatically decrypted by the protocol analyzer.
2. Intercepted raw data are exported to a comma-separated ASCII text file (see Fig. 4.14).
3. The ASCII text file is parsed in such a way that only payload coded in hexadecimal, i.e., the actual intercepted user data, will remain, as shown in Fig. 4.15.
4. Hexadecimal values in the parsed ASCII text file are used to produce the original printed binary file that was sent earlier to the printer by the Bluetooth-enabled device, i.e., the file produced contains the same information as the file that the user can produce, for example, using her printer's standard "print to file" feature.
5. Finally, the binary file produced is converted to a Portable Document Format (PDF) file, which is the final output of our Bluetooth security analysis tool.

Steps 2–5 of our BTPrinterBugging via Packet Interception Security Analysis Tool can be performed almost in real-time, especially when the exported comma-

```
1b 25 2d 31 32 33 34 35 58 40 50 4a 4c 20 4a 4f 42 20 4e 41 4d 45 3d 22 48 69 67 68 6c 79 43 6c
61 73 73 69 66 69 65 64 44 6f 63 75 6d 65 6e 74 32 2e 70 64 66 22 0a 40 50 4a 4c 20 53 45 54 20
53 54 52 49 4e 47 43 4f 44 45 53 45 54 3d 55 54 46 38 0a 40 50 4a 4c 20 43 4f 4d 4d 45 4e 54 20
22 48 50 20 43 6f 6c 6f 72 20 4c 61 73 65 72 4a 65 74 20 34 36 35 30 20 50 43 4c 20 36 20 28 42
65 6c 6b 69 6e 29 20 28 36 30 2e 33 32 2e 31 30 31 2e 34 31 29 3b 20 4d 69 63 72 6f 73 6f 66 74
20 57 69 6e 64 6f 77 73 20 58 50 20 35 2e 31 2e 32 36 30 30 2e 31 3b 20 55 6e 69 64 72 76 20 30
2e 33 2e 31 32 39 36 2e 31 22 0a 40 50 4a 4c 20 43 4f 4d 4d 45 4e 54 20 22 55 73 65 72 6e 61 6d
65 3a 20 57 4b 53 41 44 4d 49 4e 3b 20 41 70 70 20 46 69 6c 65 6e 61 6d 65 3a 20 48 69 67 68 6c
79 43 6c 61 73 73 69 66 69 65 64 44 6f 63 75 6d 65 6e 74 32 2e 70 64 66 3b 20 31 32 2d 31 2d 32
30 30 36 22 0a 40 50 4a 4c 20 53 45 54 20 4a 4f 42 41 54 54 52 3d 22 4a 6f 62 41 63 63 74 31 3d
57 4b 53 41 44 44 49 4e 4e 22 0a 40 50 4a 4c 20 53 45 54 20 4a 4f 42 41 54 54 52 3d 22 4a 6f 62 41
63 63 74 32 3d 54 4b 54 2d 4c 54 2d 4d 35 31 33 32 22 0a 40 50 4a 4c 20 53 45 54 20 4a 4f 42 41
54 54 52 3d 22 4a 6f 62 41 63 63 74 33 3d 54 4b 54 4b 62 2d 4c 54 33 31 33 33 22 0a 40 50 4a 4c
20 53 45 54 20 4a 4f 42 41 54 54 52 3d 22 4a 6f 62 41 63 63 74 34 3d 32 30 30 36 31 32 30 31 31
30 31 34 35 32 22 0a 40 50 4a 4c 20 44 4d 49 2e 4e 46 4f 20 41 53 43 49 49 48 45 58 3d 22 30 34 30
30 30 34 30 31 30 31 30 32 30 44 31 30 31 30 30 31 31 35 33 33 32 33 30 33 30 33 36 33 31 33 32 33
```

Fig. 4.15 An example of a parsed ASCII text file [2, 69]

separated ASCII text file is quite small. Even if the size of the ASCII text file is as large as two megabytes, it takes on average only 15 s on a typical Pentium laptop to produce the PDF file containing all the printed information. The attack is performed in the following way [2, 69]:

1. We discover the BD_ADDR of the non-discoverable legitimate Bluetooth-enabled device by using a Frontline FTS4BT protocol analyzer (the attacking device). The BD_ADDR of the Bluetooth-enabled printer is discovered in a few seconds via a general inquiry. If the Bluetooth link between the target devices is unencrypted, an attacker needs to know only the BD_ADDR of the Bluetooth-enabled printer, because there is no need to decrypt the intercepted information. If data decryption is needed, both BD_ADDRs must be known to the attacker (see Chap. 2).
2. We easily discover the "secret PIN code" used between the target devices by using the user-friendly name and the company_id value of the Bluetooth-enabled printer in order to download the correct user manual from the Internet.
3. We use the attacking device to intercept the traffic of the initial pairing process between the target devices.
4. We use the attacking device to intercept all the information sent to the Bluetooth-enabled printer. All intercepted data is automatically decrypted if an encrypted link is used.
5. We successfully use our BTPrinterBugging via Packet Interception Security Analysis Tool to produce a PDF file, which is the result of our practical experiment.

Figure 4.16 illustrates the result of our practical experiment in which the legitimate user printed a "Highly Classified Document" via Bluetooth. The same experiment was successfully performed with each of the four Bluetooth printer adapters using first an unencrypted link and then an encrypted link, i.e., a total of eight practical experiments were successfully performed with our BTPrinterBugging via Packet Interception Security Analysis Tool [2, 69].

A BTPrinterBugging via Packet Interception attack is very dangerous, because an attacker can steal sensitive information printed via Bluetooth. In addition, because Bluetooth is a wireless RF communication system which uses mainly omnidirectional antennas, the presence of the attacker is often unobserved. Moreover, the attacker with

Highly Classified Document   **28.11.2006**

Secret Agent: Henry Smith
Secret Agent ID#: 007
Personal Identification Number: 221277-243L
Social Security Number: 078-55-1120

This document must be kept secret! All this information is highly classified and must not leak to the wrong hands! Be very careful and destroy this message immediately after reading!

Fig. 4.16   The result of our BTPrinterBugging via Packet Interception attack [2, 69]

her attacking device can be very far away from the communicating devices, because most Bluetooth-enabled printers are class 1 devices with enhanced sensitivity levels [2, 69].

If an attacker wants to use the Bluetooth-enabled printer remotely as if it is her own, she needs to know the BD_ADDR of the printer and that of one legitimate piconet device that is using the printer. In addition, the secret PIN code that is used between these two target devices must be known to the attacker. As described earlier, in most cases the attacker can discover this secret PIN code quite quickly. Of course, the attacker must also intercept the traffic of the initial pairing process between these two devices when they meet for the first time, but as explained in Sect. 4.1, there are many different ways to arrange or force the target devices to repeat the initial pairing process. This kind of attack is normally possible only if the target devices are configured as discoverable devices. However, there are ways to find non-discoverable devices. Moreover, the attack requires that the attacker has a Bluetooth device with an adjustable BD_ADDR, i.e., the attacking device must be capable of duplicating the BD_ADDR of the legitimate device that is using the printer. Some commercially available Bluetooth protocol analyzers, such as LeCroy BTTracer/Trainer [42], support the BD_ADDR duplication feature, so this is not a problem for the attacker [2, 69].

Our Bluetooth security analysis tool, *BTPrinterBugging via Impersonation Security Analysis Tool* [2, 69], was successfully used to perform BTPrinterBugging via Impersonation attacks [2, 69] against four Bluetooth USB printer adapter models: Conceptronic, Mentor, Tecom, and Belkin. In our practical experiments we used a

USB printer with a Bluetooth USB printer adapter as the Bluetooth-enabled printer, a laptop connected to the LeCroy BTTracer/Trainer protocol analyzer [42] with one Bluetooth 1.1 compatible radio unit as the attacking device, and a Bluetooth-enabled laptop (Bluetooth 1.2 compatible) as the legitimate Bluetooth-enabled device that was using the printer. LeCroy BTTracer/Trainer v2.2 software [42], which provides CATC Scripting Language [44], was also used in our practical experiments. Our Bluetooth security analysis tool works in the following way [2, 69]:

1. We discover the BD_ADDR of the non-discoverable legitimate Bluetooth-enabled device by using a LeCroy BTTracer/Trainer protocol analyzer (the attacking device). The BD_ADDR of the Bluetooth-enabled printer is discovered in a few seconds via the general inquiry.
2. We easily discover the "secret PIN code" by using the user-friendly name and the company_id value of the Bluetooth-enabled printer. In the case of a fixed or short adjustable PIN code, an On-Line PIN Cracking attack or Off-Line PIN Recovery attack (see Sect. 4.1) can be used.
3. We use the attacking device to intercept the traffic of the initial pairing process between the target devices.
4. We use the attacking device to impersonate the legitimate piconet device by duplicating its BD_ADDR value.
5. We authenticate the attacking device with the printer using the traffic of the initial pairing process that was intercepted in step 3.
6. Finally, the attacker is capable of using the printer remotely as if it were her own: the attacking device can abuse the printer by printing funny pictures, dozens of pages of random text, and various hoax documents (see Fig. 4.17). It is also possible, for example, to print funny or lubricious pictures in order to make fun of the printer's owner or print modified "real documents" in order to mislead the legitimate users of the printer: in fact, the printer is often the best and most used source of information in a company.

As Fig. 4.17 illustrates, the attacking device first discovers the BD_ADDRs of the victim devices (see rows 1–2 and 11–12), the user-friendly names of the victim devices (see rows 3–10), and the "secret PIN code" of the printer (see row 13). The attacking device is set to require authentication and encryption for each connection with the printer (see rows 14–15). The legitimate piconet device is impersonated by duplicating its BD_ADDR value (see rows 16–18). After the successful authentication with the printer (see rows 19–25), the attacking device abuses it by printing funny pictures, dozens of pages of random text, and various hoax documents (see rows 26–28). Finally, the attacking device disconnects from the printer (see rows 29–32). The same experiment was successfully performed with each of our four Bluetooth printer adapters using first an unencrypted link and then an encrypted link, i.e., eight practical experiments were successfully performed with our BTPrinterBugging via Impersonation Security Analysis Tool. A BTPrinterBugging via Impersonation attack is typically not very dangerous, because an attacker does not steal any information from the target devices. However, this kind of attack can be very annoying if the attacker uses it to do various harmful things, as described above [2, 69].

```
(1)   Discovering the BD_ADDRs of Bluetooth Devices in Range..
(2)   The following BD_ADDRs were discovered: 000B5DA45D8C, 00027242A323
(3)   HCI_Evt> Remote_Name_Request_Complete
(4)     BD_ADDR                   : 000B5DA45D8C
(5)     Name                      : "TKT-LT-M5132"
(6)   User Friendly Name of Device 00 is: TKT-LT-M5132
(7)   HCI_Evt> Remote_Name_Request_Complete
(8)     BD_ADDR                   : 00027242A323
(9)     Name                      : "BELKIN_PRT_42A323"
(10)  User Friendly Name of Device 01 is: BELKIN_PRT_42A323
(11)  BD_ADDR of the Legitimate Bluetooth Device is: 000B5DA45D8C
(12)  BD_ADDR of the Printer is: 00027242A323
(13)  PIN Code of the Printer is: belkin
(14)  HCI_Evt> Write_Authentication_Enable_Complete
(15)  HCI_Evt> Write_Encryption_Mode_Complete
(16)  TCI_Evt> CATC_SetBdAddr_Complete
(17)    BD_ADDR                   : 000B5DA45D8C
(18)  BD_ADDR of the Legitimate Bluetooth Device is Now Successfully Duplicated!
(19)  HCI_Evt> PIN_Code_Request
(20)    PIN reply                 : belkin
(21)  HCI_Evt> Pairing_Complete
(22)    BD_ADDR                   : 00027242A323
(23)  HCI_Evt> Connection_Complete
(24)    BD_ADDR                   : 00027242A323
(25)    HCI Handle                : 0x0002
(26)  L2CAP Channel Successfully Established!
(27)  Printing Funny Pictures, Random Text and Hoax Documents..
(28)  All Printing Jobs Were Successfully Completed!
(29)  L2CAP Channel Successfully Disconnected!
(30)  HCI_Evt> Disconnection_Complete
(31)    BD_ADDR                   : 00027242A323
(32)    Reason                    : No Connection
```

Fig. 4.17 The result of our BTPrinterBugging via Impersonation attack [2, 69]

A *BTPrinterBugging via Access Denial attack* [2, 69] extends a BTPrinterBugging via Impersonation attack [2, 69], i.e., the prerequisites for these two attacks are the same. Let us assume the following attack scenario. Steps 1–5 in this attack are the same as in a BTPrinterBugging via Impersonation attack. Finally (step 6), the attacker is able to deny the legitimate piconet users access to the printer by making repeated successful connection establishments to the printer, i.e., the attacker makes sure that the printer is always busy and therefore unable to service other devices [2, 69].

Our Bluetooth security analysis tool, the *BTPrinterBugging via Access Denial Security Analysis Tool* [2, 69], was successfully used to perform BTPrinterBugging via Access Denial attacks [2, 69] against four Bluetooth USB printer adapter models: Conceptronic, Mentor, Tecom, and Belkin. All other hardware and software used in this practical experiment were the same as in the BTPrinterBugging via Impersonation attack. Our Bluetooth security analysis tool works in the following way. Steps 1–5 are the same as in a BTPrinterBugging via Impersonation attack. In step 6, the attacking device successfully establishes connection with the printer, waits until the printer automatically performs the disconnection, and makes similar repeated successful connection establishments. In this way, the attacking device denies the legitimate piconet devices access to the printer [2, 69].

The same experiment was successfully performed with each of our four Bluetooth printer adapters using first an unencrypted link and then an encrypted link, i.e., eight practical experiments were successfully performed using our BTPrinterBugging via Access Denial Security Analysis Tool [2, 69].

Some Bluetooth-enabled printers never perform automatic disconnection, so an attacker can establish basic ACL links with them which are never disconnected. In this way, the attacker denies the legitimate piconet devices access to the printers. Some Bluetooth-enabled printers also support multiple connections at the same time. In this case, the attacker can simply use several Bluetooth radio units to open multiple ACL links with the printer adapter. Another possibility is to use one Bluetooth radio unit, but open multiple L2CAP channels with the printer in order to reserve all resources for the attacker [2, 69].

We also implemented another Bluetooth security analysis tool, *BTPrinterBugging via Access Denial Security Analysis Tool II* [2, 69], which works in Linux environments. It requires a BlueZ protocol stack [37] and at least one Bluetooth USB dongle to work, i.e., an expensive Bluetooth protocol analyzer is not required. This security analysis tool does not require any tricks that only sophisticated protocol analyzers can do. The only required information is the BD_ADDR of the Bluetooth-enabled printer, i.e., the attacker does not even have to know the secret PIN code of the Bluetooth-enabled printer. Our BTPrinterBugging via Access Denial Security Analysis Tool II is capable of keeping Bluetooth-enabled printers busy all the time by making repeated connection attempts to them. In this way, the attacking device very easily denies the legitimate piconet devices access to all Bluetooth-enabled printers. Our security analysis tool supports the automatic discovery and recognition of Bluetooth-enabled printers in range. Another possibility is to select Bluetooth-enabled printers manually from the list of discovered devices. If several Bluetooth-enabled printers are detected, several Bluetooth USB dongles will also automatically be used in parallel, i.e., one Bluetooth USB dongle is used for each detected Bluetooth-enabled printer. This security analysis tool can also be used to perform *DoS attacks* against any other kinds of Bluetooth devices, because the only information required is the BD_ADDR of the target device [2, 69].

Figure 4.18 illustrates our BTPrinterBugging via Access Denial Security Analysis Tool II in action (see rows 1–34). When the attacker runs the BTPrinterBugging via Access Denial Security Analysis Tool II in Linux (see row 1), she can either manually select Bluetooth-enabled printers from the list of discovered devices (see rows 2–3) or alternatively use automatic discovery of Bluetooth-enabled printers (see rows 4–5). In this practical experiment, the attacker chooses the automatic discovery option (see rows 6–15). Because four Bluetooth-enabled printers are detected, four Bluetooth USB dongles are used in parallel (see row 16). Finally, the attacking device simultaneously denies all legitimate printer users access to all four Bluetooth-enabled printers (see rows 17–34) [2, 69].

Some Bluetooth-enabled printers allow users to create basic ACL links without authentication, i.e., no PIN code is required prior to accepting the connection establishment between the Bluetooth-enabled printer and a remote Bluetooth-enabled device. Therefore, an attacker can very easily reserve all the resources of such printers by using several Bluetooth USB dongles to establish multiple ACL links. BTPrinterBugging via Access Denial attacks are typically not very dangerous, because an attacker does not steal any information from the target devices. However, these kinds of attacks can be very annoying if the attacker uses them non-stop in order to perma-

```
(1)  # ./BTPrinterBuggingViaAccessDenial2
(2)  1. Select Bluetooth-enabled Printers Manually From the List of
(3)     Discovered Devices.
(4)  2. Automatically Both Discover And Decide What Devices Are
(5)     Bluetooth-enabled Printers.
(6)  2
(7)  Scanning ...
(8)          00:02:72:42:A3:23      BELKIN_PRT_42A323
(9)          00:04:61:10:05:A0      DESKTOP_PC-FC6
(10)         00:04:61:83:9A:26      Printer Adapter 839A26
(11)         00:0B:0D:10:29:A1      BT printer
(12)         00:0B:5D:A4:5D:8C      TKT-LT-M5132
(13)         00:03:C9:3C:33:32      Sitecom CN-505
(14) Bluetooth Devices Found: 6
(15) Bluetooth-enabled Printers Detected: 4
(16) Using 4 Bluetooth USB Dongles in Parallel.
(17) Simultaneously Denying Access From All Legitimate Printer Users
(18) to All 4 Bluetooth-enabled Printers:
(19)  Connecting to Printer Adapter #1 (00:02:72:42:A3:23) (attempt #1)
(20)  Connecting to Printer Adapter #2 (00:04:61:83:9A:26) (attempt #1)
(21)  Connecting to Printer Adapter #3 (00:0B:0D:10:29:A1) (attempt #1)
(22)  Connecting to Printer Adapter #4 (00:03:C9:3C:33:32) (attempt #1)
(23)   Can't Create Connection to Printer Adapter #1: Connection Timed Out
(24)   Can't Create Connection to Printer Adapter #2: Connection Timed Out
(25)   Can't Create Connection to Printer Adapter #3: Connection Timed Out
(26)   Can't Create Connection to Printer Adapter #4: Connection Timed Out
(27)  Connecting to Printer Adapter #1 (00:02:72:42:A3:23) (attempt #2)
(28)  Connecting to Printer Adapter #2 (00:04:61:83:9A:26) (attempt #2)
(29)  Connecting to Printer Adapter #3 (00:0B:0D:10:29:A1) (attempt #2)
(30)  Connecting to Printer Adapter #4 (00:03:C9:3C:33:32) (attempt #2)
(31)   Can't Create Connection to Printer Adapter #1: Connection Timed Out
(32)   Can't Create Connection to Printer Adapter #2: Connection Timed Out
(33)   Can't Create Connection to Printer Adapter #3: Connection Timed Out
(34)   Can't Create Connection to Printer Adapter #4: Connection Timed Out
```

Fig. 4.18 The BTPrinterBugging via Access Denial Security Analysis Tool II in action [2, 69]

nently deny the legitimate piconet users access to the printers. Moreover, the attacker can use these kinds of attacks to mislead the target devices in such a way that they delete previously stored link keys so that the initial pairing process is restarted [2, 69].

Bluetooth worms and viruses have been often mentioned in the mass media as well as in research articles, for example, in [47, 75, 76], because there are several Bluetooth worms and viruses, such as Cabir [77], Skulls.D [78], and Lasco.A [79], which use Bluetooth-enabled mobile phones to infect other Bluetooth mobile phones [2].

Cabir [77] (also referred to as *SymbOS/Cabir.A, EPOC/Cabir.A, Worm. Symbian.Cabir.a,* or *Caribe virus*) is a Bluetooth worm running in Symbian mobile phones which support the Series 60 platform. It arrives via Bluetooth in a target mobile phone's messaging inbox as a caribe.sis file, which contains caribe.app (main executable), flo.mdl (system recognizer), and caribe.rsc (resource file). When the user opens the caribe.sis file and chooses to install it, the worm activates and immediately starts searching for new Bluetooth devices to infect. When another Bluetooth device is found, Cabir will start sending the caribe.sis file to it. However, the file will not arrive automatically in the target device, because the user has to answer "yes" to the transfer question while the infected device, i.e., the attacking device, is still in range. It is worth noting that Cabir only spreads itself, i.e., it is not designed to do any harmful things such as erasing a target device's files [2, 77].

Skulls.D [78] (also referred to as *SymbOS/Skulls.D*) is a malicious Symbian Installation System (SIS) file trojan that pretends to be Macromedia Flash player for Symbian mobile phones which support the Series 60 platform. It arrives in the target mobile phone via Bluetooth in a similar way to Cabir. When the user opens the SIS file and chooses to install it, the SymbOS/Cabir.M worm, i.e., the variation of the Cabir worm described earlier, will be installed in the target mobile phone, both the system applications and third party applications that are needed to disinfect viruses and worms will be disabled, and animation showing a flashing skull picture will also be displayed on the background of the target device's display during every application that the user is trying to use. When the worm is activated, it immediately starts searching for new Bluetooth devices to infect [2, 78].

Lasco.A [79] (also referred to as *SymbOS/Lasco.A* or *EPOC/Lasco.A*) is a Bluetooth worm and a SIS-file-infecting virus running in Symbian mobile phones which support the Series 60 platform. It arrives in the target mobile phone via Bluetooth in a similar way to Cabir and Skulls.D. When the user opens the velasco.sis file and chooses to install it, the worm activates and immediately starts searching for new Bluetooth devices to infect. In addition to sending itself via Bluetooth, it is also capable of inserting itself into other SIS files in the target device. Therefore, if infected SIS files are copied to another device and installed, installation of Lasco.A will also start [2, 79].

In January 2005, Brazilian software developer Marcos Velasco released all source codes of his Lasco.A worm/virus on his homepage [80], but later removed them. Lasco.A sources can still be downloaded from many Brazilian file servers. It means that practically anyone can now write their own Bluetooth viruses just by modifying Lasco.A sources. In addition, a mobile phone infected with Cabir, Skulls.D, or Lasco.A will try to infect other Bluetooth-enabled devices even if the user tries to disable Bluetooth from the device's settings. Moreover, Bluetooth functionality in Series 60 mobile phones is independent from the GSM side. Therefore, if the infected mobile phone is rebooted, the virus/worm will try to spread itself even if the user does not enter the PIN code [2, 79, 80].

Normally, Bluetooth worms and viruses require that the user accepts their transfer and installation in the target device. Moreover, the target device has to be discoverable. Bluetooth worms and viruses can be very dangerous if the target device is vulnerable to BlueBugging, because in that way an attacker can slip in a virus or worm without alerting the user. Therefore, if a user has a vulnerable Bluetooth device, its firmware/software should be updated as soon as possible. Furthermore, the user should not install any unknown software in her Bluetooth device and she should use antivirus/firewall software when possible. It is also expected that attackers will exploit the techniques for finding hidden Bluetooth devices in an average of one minute [30, 31] in order to spread viruses and worms more efficiently [2].

Chapter 5
MITM Attacks on Bluetooth

Our MITM attacks on SSP, *BT-Niño-MITM attack* [2, 20], *BT-SSP-OOB-MITM attack* [2, 22], *BT-SSP-Printer-MITM attack* [2, 21], and *BT-SSP-HS/HF-MITM attack* [2, 22] as well as a *SSP MITM attack of Suomalainen et al.* [19], are described in Sect. 5.1. Section 5.2 provides a literature-review-based comparative analysis of the existing MITM attacks on Bluetooth over the past ten years (2001–2011), including our MITM attacks on Bluetooth SSP.

5.1 MITM Attacks on SSP

We call our first attack a *BT-Niño-MITM attack* [2, 20] (also referred to as a *Bluetooth—No Input, No Output—Man-In-The-Middle attack*). In the attack we exploit the fact that the devices must exchange information about their IO capabilities during the first phase of the SSP (see Chap. 2). The exchange is done over an unauthenticated channel, and an attacker that controls this channel can therefore modify the information about capabilities and force the devices to use the association model of her choice. In our attack, the devices are forced to use the Just Works association model, which does not provide protection against MITM attacks. The MITM uses two separate Bluetooth devices with adjustable BD_ADDRs for the attack. Such devices are readily available on the market. The MITM clones the BD_ADDRs and user-friendly names of the victim devices in order to impersonate them more plausibly. The main features of the attack are depicted in Fig. 5.1. We next describe three general scenarios for the attack [2, 20].

In the first scenario, the MITM first disrupts (jams) the PHY by hopping along with the victim devices and sending random data in every timeslot. Another possibility is to jam the entire 2.4 GHz band altogether by using a wideband signal. In this way, the MITM shuts down all piconets within the range of susceptibility and there is no need to use a Bluetooth chipset to generate hopping patterns. Finally, the frustrated user thinks that something is wrong with her Bluetooth devices and deletes previously

K. Haataja et al., *Bluetooth Security Attacks*, SpringerBriefs in Computer Science,
DOI: 10.1007/978-3-642-40646-1_5, © The Author(s) 2013

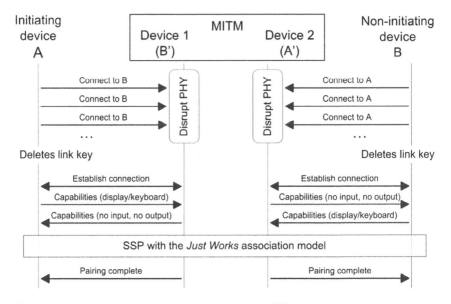

Fig. 5.1 The main features of a BT-Niño-MITM attack [2, 20]

stored link keys. Then the user initiates a new pairing process by using SSP, and the MITM can forge messages exchanged during the IO capabilities exchange phase. When the Just Works association model has been forced into use, the attack continues as illustrated in Fig. 5.2. Most of the notations used in Fig. 5.2 have been explained earlier (see Table 2.2 in Chap. 2) and the rest are self-explanatory [2, 20].

It is worth noting that in this first scenario two victim devices have already performed the initial pairing, including the capabilities exchange, so link keys are saved on the devices for use in subsequent connections, i.e., the victim devices normally use SSP without capabilities exchange (see Chap. 2). Other scenarios, where victim devices have never met before, are easier for the MITM, because in those cases the first phase of the attack (disrupting the PHY) can be skipped. There can be two different scenarios for such devices [2, 20]:

1. *The victim device (A or B) initiates SSP*: In this scenario, the MITM waits until A or B initiates SSP. Then the attack proceeds as illustrated in Figs. 5.1 and 5.2.
2. *The MITM (A' and B') initiates SSP*: In this scenario, the MITM first initiates SSP with the victim devices. Then the attack proceeds as illustrated in Figs. 5.1 and 5.2.

Depending on the situation, the MITM can use any of these three attack scenarios. The applicability of a certain attack scenario depends on the implementation of the victim devices. Moreover, it may be possible to perform SSP without asking the user to accept the connection, because this confirmation request is optional in the Bluetooth specification [1, 2, 20].

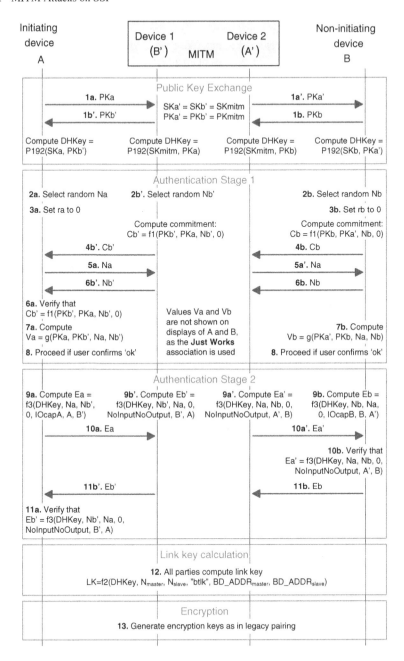

Fig. 5.2 Pairing details of a BT-Niño-MITM attack [2, 20]

Suomalainen et al. [19] have performed a comparative analysis of Bluetooth SSP, Wireless Fidelity (Wi-Fi) Protected Setup, Wireless USB Association Models, and HomePlugAV security modes. They present an attack against SSP similar to our BT-Niño-MITM attack. In their attack the MITM prompts one device to use the normal Numeric Comparison association model, while forcing the other device to use the insecure Just Works association model. This leads to one of the devices (the one which uses the Numeric Comparison association model) treating the resulting link key as authenticated, and it might choose to trust it even for an application which requires a high level of security. However, this attack looks somewhat suspicious from the point of view of the user: one of the devices asks the user to compare the integrity checksums, while the other device does not display any numbers. In the tests performed by Suomalainen et al. [19], only six users out of 40 accepted the pairing on both devices. *Compared with the SSP MITM attack of Suomalainen et al. [19], our BT-Niño-MITM attack looks less dubious*: indeed, the user is only asked to confirm the pairing on both devices by pressing a button. In addition, since this confirmation request is optional in the Bluetooth specification [1], some manufacturers might choose to skip it in order to improve usability. Moreover, as the MITM in our attack uses two Bluetooth devices with BD_ADDRs and user-friendly names equal to those of the victim devices, the user becomes even more confident that the pairing is proceeding correctly and securely. It is also worth noting that by using two MITM devices, SSP can be performed at the same time with both victim devices and it also ends at the same time with both victim devices, thus making the user even more confident [2, 20].

We call our second attack a *BT-SSP-OOB-MITM attack* [2, 22] (also referred to as a *Bluetooth—Secure Simple Pairing—Out-Of-Band—Man-In-The-Middle attack*). The attack requires that the attacker can somehow see the victim devices, i.e., there must be some kind of visual contact (for example, a hidden video camera or direct line of sight) to the victim devices. In the attack legitimate users are misled to select a less secure option instead of using a more secure OOB channel, for example, USB cable, Infrared Data Association (IrDA), or NFC. The attack works against any two OOB-capable Bluetooth devices that support SSP. The main features of the attack are depicted in Fig. 5.3. We next describe a general scenario for the attack [2, 22].

Our attack works in the following way [2, 22]:

1. *The MITM impersonates the legitimate devices*: The MITM uses two Bluetooth devices with BD_ADDRs and Bluetooth names equal to those of the victim devices, i.e., the user becomes confident that the pairing is proceeding correctly and securely.
2. *The MITM acts just before the legitimate user*: The attacker has visual contact to the victim devices and he notices that the user is about to start SSP with the OOB association model. The MITM acts just before the legitimate user and establishes connections to both victim devices in order to start the capabilities exchange.
3. *The MITM forces the Just Works association model into use*: The MITM forces the victim devices to use the Just Works association model and after that the attack continues as illustrated in Fig. 5.2.

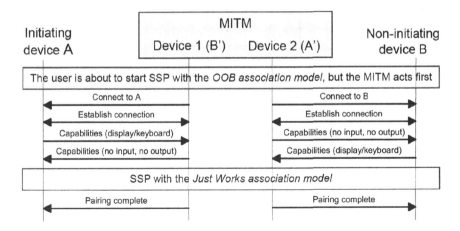

Fig. 5.3 The main features of a BT-SSP-OOB-MITM attack [2, 22]

We call our third attack a *BT-SSP-Printer-MITM attack* [2, 21] (also referred to as a *Bluetooth—Secure Simple Pairing—Printer—Man-In-The-Middle attack*). In this attack we exploit the fact that almost all Bluetooth-enabled printers that support SSP (especially those connected using Bluetooth USB printer adapters) will use the Just Works association model in order to make printing user-friendly. It is not likely that users will be required to press any printer buttons just to accept the connection establishment in the initial pairing process of SSP. Therefore, the Just Works association model seems to be the most logical choice for SSP-enabled printers. Our attack works even against SSP-enabled printers that provide MITM protection via the Numeric Comparison, the Passkey Entry, or the OOB association model, because victim devices can be forced to use any association model that the attacker chooses [2, 19–23]. The main features of the attack are depicted in Fig. 5.4. We next describe two scenarios for the attack [2, 21].

We assume in the first scenario that the victim devices use the Just Works association model. Our attack works in the following way [2, 21]:

1. *The MITM disrupts the PHY until the frustrated user deletes previously stored link keys.*
2. *The MITM impersonates the legitimate printer:* Since the user has deleted the previously stored link keys, she will initiate a new pairing process through SSP. The SSP pairing details are illustrated in Fig. 5.2. It is worth noting that the user has deleted all information about the legitimate printer, including its BD_ADDR, so the MITM is not even required to clone the BD_ADDR of the legitimate printer in order to impersonate it. Now the MITM only clones the user-friendly name of the legitimate SSP-enabled printer to impersonate it. Moreover, the MITM must be able to disrupt the legitimate printer in such a way that it cannot communicate with other legitimate Bluetooth devices. Therefore, when the user seeks available Bluetooth printers in range, the only printer that is found will be the MITM with a different BD_ADDR but the same familiar user-friendly name. It is very likely that

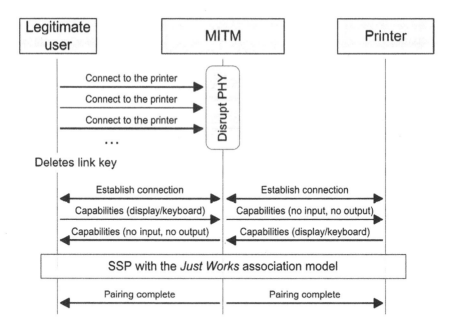

Fig. 5.4 The main features of a BT-SSP-Printer-MITM attack [2, 21]

the user will not notice anything strange, because BD_ADDRs are much harder to remember than user-friendly names. Therefore, the user will most likely choose the "MITM printer" that looks familiar to her. In this way, the MITM has replaced the legitimate printer in the Bluetooth network with the "MITM printer" with a different BD_ADDR. It is worth noting that by using a BD_ADDR different from that of the legitimate printer, the MITM can also eliminate any BD_ADDR collisions that may occur, i.e., the attack works more reliably and plausibly.

3. *The MITM intercepts all data*: When the legitimate Bluetooth devices print via a Bluetooth connection, the MITM captures (receives) all data and is also capable of decrypting it if encryption is used.

4. *The MITM relays the captured data to the legitimate printer*: In this way, everything seems to work normally from the user's point of view, because all documents are printed without any problems.

We assume in the second scenario that the victim devices use the Numeric Comparison, the Passkey Entry, or the OOB association model. This attack works in the same way as our first attack scenario except that one additional phase is required: *the legitimate devices must be forced to use the Just Works association model.* Note that since our two attack scenarios are designed against Bluetooth 2.1+EDR/3.0+HS/4.0 (SSP-enabled) printers, a MITM device is required between the victim devices for the attacks to work. Attacks against Bluetooth 2.0+EDR and earlier printers are easier in practice, because the MITM device is not required. Such attack scenarios and their practical implementations were described in Sect. 4.4 [2, 21].

We call our fourth attack a *BT-SSP-HS/HF-MITM attack* [2, 22] (also referred to as a *Bluetooth—Secure Simple Pairing—Headset/Hands-Free—Man-In-The-Middle attack*). In the attack we exploit the fact that almost all Bluetooth-enabled headsets and hands-free devices that support SSP will use the Just Works association model in order to make the pairing process user-friendly. Our attack works even against SSP-enabled headsets and hands-free devices that provide MITM protection via the Numeric Comparison, the Passkey Entry, or the OOB association model, because victim devices can be forced to use the Just Works association model [2, 19–23]. We next describe a scenario for the attack [2, 22].

Our attack works in the following way [2, 22]:

1. *The MITM impersonates the legitimate devices by using two Bluetooth devices with BD_ADDRs and Bluetooth names equal to those of the victim devices.*
2. *The MITM disrupts the PHY until the frustrated user deletes previously stored link keys.*
3. *The legitimate user initiates a new pairing process*: If the victim devices use the Just Works association model, the attacker does not have to forge the messages exchanged during the IO capabilities exchange phase, i.e., it is enough just to relay messages from one victim to the other. In the case when victim devices are using the Numeric Comparison, the Passkey Entry, or the OOB association model, the attacker must force the Just Works association model into use. In the case when the OOB association model is used by the victim devices, the attacker must have some kind of visual contact to the victim devices in order to force the Just Works association model into use. After the capabilities exchange phase, the attack continues as illustrated in Fig. 5.2.

After a successful BT-Niño-MITM attack, BT-SSP-OOB-MITM attack, BT-SSP-Printer-MITM attack, BT-SSP-HS/HF-MITM attack, or SSP MITM attack of Suomalainen et al., the MITM can intercept and modify all data exchanged between the victim devices, and even use certain services that the victim devices offer [2, 19-23].

Because it is difficult to combine high levels of security with good usability, other studies of SSP have concentrated mostly on analyzing and improving its usability. Uzun et al. [81] have analyzed different ways of prompting the user to perform the comparison of integrity checksums or to enter passkeys. They have also provided guidelines for designing the user interface and decreasing the number of fatal errors, thus improving both the usability and security of SSP [2, 20].

5.2 Comparative Analysis of Bluetooth MITM Attacks

The first MITM attack on Bluetooth was devised in 2001 by Jakobsson and Wetzel [13]: the attack was also the world's first Bluetooth security attack and thus *the year 2011 was a major 10-year milestone in the history of Bluetooth security attacks*—this milestone was also a big motivator in writing this rather extensive book.

Even though the attack was devised for version 1.0B of the Bluetooth standard, it works also with all Bluetooth versions up to 2.0+EDR, because no major security improvements were implemented in those Bluetooth specifications. The attack assumes that the link key used by the two victim devices is known to the attacker. The authors also showed how to obtain the link key using an Off-Line PIN Recovery attack (see Sect. 4.1). The MITM attack requires that both devices have their security levels set to public or private (see Chap. 2), i.e., both victim devices must be connectable. In the attack, the BD_ADDRs of the attacker's devices must be cloned to equal the addresses of the victim devices. Moreover, to prevent the jamming of the communication channel, the victim devices must be both masters or both slaves (in two different piconets). In this case they transmit in an unsynchronized manner and cannot see each other's messages, while communicating with the attacker. After establishing a connection to both victims, the attacker sets up two new link keys [2, 13, 21, 23].

Kügler [14] further improves the attack of Jakobsson and Wetzel. By manipulating the clock settings, the attacker forces both victim devices to use the same channel hopping sequence but different clocks. In this way, the victim devices are unsynchronized and can see only the messages the attacker sends them. Moreover, Kügler [14] shows how a MITM attack can be performed during the paging procedure. The attacker responds to the page request of the master victim faster than the slave victim and restarts the paging procedure with the slave using a different clock. The master and slave use the same channel hopping sequence, but a different offset in this sequence. The attack also works in case both victim devices send and receive data packets over an encrypted link. Even though the Initialization Vector (IV) used for encryption depends on the clock, the last bit of the clock is unused. Therefore, the attacker can flip this last bit, forcing the victims to use clocks which have a difference of approximately 11.65 h. Although the integrity of the data is protected with CRCs which are appended to the plaintext prior to encryption, the attacker can manipulate the intercepted ciphertext. After modifying the ciphertext in a certain way, the attacker updates the CRC bits (see Ref. [82] for details): the integrity checks performed by the victims do not detect the modification. It must be noted, however, that the attacker does not have much time for manipulating the transmitted data [2, 14, 21, 23].

As we discussed in Sect. 4.2, *Reflection attacks* [15] (also referred to as *Relay attacks*) can also be seen as a type of a MITM attack against authentication, but not encryption. The only information needed is the BD_ADDRs of the victim devices. In a *One-Sided Reflection attack* only one victim device is impersonated, while in a *Two-Sided Reflection attack* both victim devices are impersonated (see Sect. 4.2 for more information). During the paging procedure, the attacker responds to the request of the first victim device (A), and initiates a connection to the second victim device (B), posing as A. If the victim devices can hear each other, the mechanisms described in [14] can be used to achieve this. After this, the attacks work on the Link Manager Protocol (LMP) layer of Bluetooth. The messages of the protocol are simply relayed by the attacker's devices. In the case of a one-sided attack, only part of the messages must be relayed, and the connection to A is dropped when the attacker has

Table 5.1 MITM attacks on Bluetooth: summary and comparison [2, 21, 23]

Attack	[13]:	[14]:	[15]:	[19]:	[20–22]:
Bluetooth versions:	1.0 – 2.0 + EDR	1.0 – 2.0 + EDR	1.0 – 2.0 + EDR	2.1 + EDR – 4.0	2.1 + EDR – 4.0
Attack goals:	Impersonation, modification	Impersonation, modification	Impersonation	Impersonation, modification	Impersonation, modification
Attacking devices:	2	2	2	$2+1^a$	$2+1^a$
Devices attacked:	Connectable	Connectable or non-connectable	Connectable or non-connectable	Connectable or non-connectable	Connectable or non-connectable
Distances:	Any^b	Any^b	$Any^{b,c}$	Any^b	Any^b
Detection:	By $user^d$	$None^e$	By $devices^f$	By $user^g$	By $user^h$
Main counter-measure:	Security $policies^i$	Integrity $checks^j$	Detecting the delays	At the user interface level	At the user interface $level^k$

[a] A jamming device is also required.

[b] Actual distance is limited by the speed of the link between the attacker's devices. The attacker must use two Bluetooth adapters.

[c] The victim devices must be out of each other's range.

[d] The user enters the PIN to renegotiate.

[e] The attack remains undetected.

[f] There are delays in getting the LMP authentication response.

[g] One of the devices asks the user to compare numbers, while the other one does not.

[h] No Numeric Comparison is used although both devices have displays and keyboards.

[i] Security policies protecting against MITM attacks are proposed in Sect. 6.3.

[j] Cryptographic integrity checks of packets should be used.

[k] Modifications to SSP specification are proposed in Sect. 6.4.

impersonated it to B. The attacker can successfully perform authentication by using reflection attacks, but she cannot continue the attack if the target devices encrypt their communication. By combining reflection attacks with a known secret PIN code, link key, or encryption key, the attacker can both impersonate the victim devices and decrypt the information transferred between them. Victim devices can detect the attack by noticing a considerable increase in the latency of the response to the authentication challenge, caused by relaying. This countermeasure is not described in the standards, and it is up to the discretion of manufacturers whether to provide it [2, 14, 15, 21, 23].

Versions 2.1+EDR, 3.0+HS, and 4.0 of Bluetooth provide protection against the MITM attacks described above, by the means of SSP as described in Chap. 2. However, it has been shown that MITM attacks against SSP-enabled Bluetooth devices are also possible by forcing victim devices to use the Just Works association model [2, 19–23] (see Sect. 5.1). Thus, the attacker can bypass all security checks which would normally be in place. The association is then unauthenticated: the devices are aware of this fact, but it depends on the manufacturer how they react to this. If the victim devices have already established a link key, the attacker can use jamming to disrupt the communication and then she can initiate the connection with both devices under a chosen association model. As a result, the attacker learns the link key used by the devices and she can intercept all data transmitted between the devices. In Table 5.1 we summarize the properties of the MITM attacks outlined in this section [2, 21, 23].

It is interesting to note the connection of MITM attacks to other developments in Bluetooth security analysis. For instance, at the time when most of the MITM attacks were introduced, implementing them was not an easy task, as there were no devices with adjustable BD_ADDRs, except sophisticated and expensive protocol analyzers. Now the situation has changed: Bluetooth devices with an adjustable BD_ADDR are readily available and techniques for finding hidden (non-discoverable) Bluetooth devices have been invented (see Sect. 4.1). Therefore, the danger of MITM attacks has recently increased. Indeed, at least one of the proposed MITM attacks against Bluetooth SSP has already been implemented and even mounted in practice [2, 21, 23, 24].

Chapter 6
Countermeasures

Various Bluetooth security attacks can be prevented and stopped by monitoring communication to discover such attacks: thus, we have proposed an *Intrusion Detection and Prevention System* [2, 83] for Bluetooth networks to prevent and stop attacks in progress (see Sect. 6.1). Furthermore, we feel that the use of *RF fingerprints* (also referred to as *RF signatures*) [84–90] could be the future of secure Bluetooth communications: therefore, we have proposed an *RF Fingerprint-Based Security Solution for SSP* [90] (see Sect. 6.2). Finally, Sect. 6.3 provides countermeasures for Bluetooth devices up to 2.0+EDR, while Sect. 6.4 provides countermeasures for SSP-enabled Bluetooth devices.

6.1 Intrusion Detection and Prevention System

Our *Intrusion Detection and Prevention System* [2, 83] for Bluetooth networks is based on the set of rules that are used to identify strange communication behavior of Bluetooth devices. Based on the strange communication behavior of Bluetooth devices which are undergoing various security attacks, we defined a set of rules to help identify attacks in progress [2, 83]:

1. *Unusually many repeated failed authentication attempts:* This may indicate that an attacker is using an On-Line PIN Cracking attack (see Sect. 4.4) to discover the secret PIN code of the victim device.
2. *Unusually many repeated successful authentications and disconnections:* This may indicate that an attacker is performing a DoS attack (see Sect. 4.3).
3. *Unusually many NAK transmissions:* This may indicate that an attacker is performing a Big NAK attack (see Sect. 4.3), thus putting the victim device on an endless retransmission loop.
4. *Unusually long delays:* This may indicate that a MITM is between the communicating parties (see Chap. 5).

K. Haataja et al., *Bluetooth Security Attacks*, SpringerBriefs in Computer Science,
DOI: 10.1007/978-3-642-40646-1_6, © The Author(s) 2013

5. *Unusually many repeated POLL packets:* This may indicate that an attacker is keeping victim devices busy so that they will not go into sleep or low-power mode (see Chap. 7).
6. *Unusually high BER:* This may indicate that an attacker is disrupting the PHY (see Sect. 4.3).
7. *Unusually heavy traffic between two communicating parties:* This may indicate that an attacker is performing a Battery Exhaustion attack (see Sect. 4.3).
8. *Sudden increase in transmit powers:* This may indicate that the attacker is using a stronger RF signal to displace the active piconet device via an Exploitation of a stronger RF signal attack (see Sect. 4.2).
9. *Two identical BD_ADDRs in the range of vulnerability:* This may indicate that an attacker is using a BD_ADDR Duplication attack (see Sect. 4.3) to deny the legitimate piconet devices access to services. Another possibility is that an attacker is performing an impersonation attack (for example, a BTPrinterBugging via Impersonation attack; see Sect. 4.4) to mislead the legitimate piconet devices.
10. *An HV1 SCO link established with the piconet master when another type of SCO or eSCO link could also have been used:* This may indicate that an attacker is performing a SCO/eSCO attack (see Sect. 4.3) to reserve all piconet resources so that the legitimate piconet devices do not receive service within a reasonable time.
11. *An L2CAP level request for the highest possible data rate or the smallest possible latency:* If such a request is accepted, all throughput is reserved for the attacker and the legitimate piconet devices do not receive service within a reasonable time, i.e., the attacker is performing an L2CAP Guaranteed Service attack (see Sect. 4.3).
12. *Surprising connection attempts and data transfer requests from unknown Bluetooth devices:* This may indicate that a Bluetooth virus or worm (see Sect. 4.4) is trying to infect legitimate piconet devices.
13. *A Bluetooth device requests that the length of an encryption key must be shorter than 128 bits:* This may indicate that an attack against the Bluetooth encryption is in progress (see Sect. 3.2).
14. *An RF signature mismatch:* This indicates that some kind of attack, such as an impersonation attack, is in progress. Every transmitter has a unique RF signature [2, 12, 84] which can be used to differentiate legitimate devices from those that have alien RF signatures, i.e., a sample RF signature is needed from each legitimate device in order to detect alien RF signatures (see Sect. 6.2).
15. *SSP's Just Works association model activated between devices that could use a more secure option (for example, Numeric Comparison or OOB):* This may indicate that a MITM is between the communicating parties (see Chap. 5).

In order to mitigate various Bluetooth security attacks, we have proposed a scheme consisting of two parts: an *Intrusion Detection System* and an *Intrusion Prevention System* [2, 83]. In our scheme a commercially available Bluetooth protocol analyzer, such as a LeCroy BTTracer/Trainer [42] as well as its v2.2 software [42] (or later) providing CATC Scripting Language [44], equipped with signal processing capabilities (some additional signal processing hardware is required) takes care of the intrusion detection part. When an intrusion is detected, the protocol analyzer immediately informs the network administrator that the Bluetooth network is under attack. This is *manual administrative intrusion prevention*, which can be used in all cases regardless of the capabilities of the legitimate Bluetooth devices [2, 83].

The second part of our system, the *Intrusion Prevention System*, is a small program that runs on all legitimate Bluetooth devices that allow programs to be installed, i.e., at least all PCs, laptops, and mobile phones should be supported. This is *automatic intrusion prevention*. It requires that all legitimate Bluetooth devices must run this special program in order to receive warning messages from our Intrusion Detection System. When a warning message is received, devices that are under attack perform automatic disconnection and refuse any further Bluetooth connections for a predetermined time. The Intrusion Detection System also sends enough information (BD_ADDR, device capabilities information, the user-friendly name of the device, RF signature information, and so on) to the Intrusion Prevention System so that further connections from the same origin can be refused immediately by the Intrusion Prevention System [2, 83].

Our Intrusion Detection System works in the following way [2, 83]:

1. *A Bluetooth protocol analyzer monitors Bluetooth communications of the legitimate piconet devices non-stop:* The protocol analyzer has all the legitimate BD_ADDR values that are allowed to communicate within the piconet and also other useful information about such devices (for example, device capabilities information, the user-friendly name of the device, and RF signature information).
2. *When an intrusion is detected, either manual administrative intrusion prevention or automatic intrusion prevention is applied:* We strongly recommend that automatic intrusion prevention should be implemented. However, if automatic intrusion prevention is not implemented, at least the Bluetooth network administrator is alerted immediately.

Our automatic Intrusion Prevention System works in the following way [2, 83]:

1. *The Intrusion Prevention System receives a warning message from the Intrusion Detection System:* When a warning message is received, automatic disconnection is performed and further Bluetooth connections are refused for a predetermined time.
2. *The Intrusion Prevention System also receives enough information to prevent further attacks from the same origin:* We recommend that at least the following information about the attacking device should be received from the Intrusion Detection System and stored in the database: the BD_ADDR value, device capabilities information, the user-friendly name of the device, and RF signature information.

It is difficult to create an intrusion detection and prevention system that caters for all possible types of security attacks, as the security of Bluetooth is likely to be limited by the capabilities of the least powerful or the least secure device type: in fact, most Bluetooth security attacks are based exactly on this problem. In general, security attacks are hard to prevent in wireless networks, especially when no intrusion detection and prevention system is used. Our proposal is intended to help Bluetooth network administrators and Bluetooth device manufacturers to implement efficient Bluetooth intrusion detection and prevention systems [2, 83].

6.2 RF Fingerprint-Based Security Solution for SSP

Even if devices are produced by the same manufacturer using the same components, there are differences between signals sent by these RF devices. These small differences are the result of variations in the electronical components of a device. Thus, RF devices can be identified and differentiated from each other. Variances are most evident when a device is being activated or when it tries to access a network, because then there exists a short transient phase in the signal. This transient phase lasts only 2–10 ms. In the transient phase, there often occur significant changes in frequency, amplitude, and phase: the RF fingerprint can be formed from this particular part of the signal. Figure 6.1 illustrates a typical Bluetooth signal [84-90].

Since every transmitter has a unique RF fingerprint, it can be used to differentiate legitimate devices from devices that have alien RF fingerprints. For this purpose, a sample RF fingerprint is needed from each legitimate device in order to detect alien RF fingerprints. Wireless devices, such as Bluetooth devices, can be equipped with signal processing capabilities to check every RF fingerprint before accepting

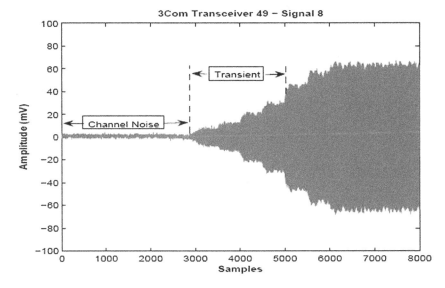

Fig. 6.1 A typical Bluetooth signal [89, 90]

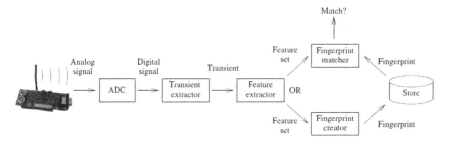

Fig. 6.2 RF-fingerprinting process [90, 91]

any connections. We feel that RF fingerprints could play a major role for improving the security of Bluetooth, because RF fingerprints are extremely hard to duplicate (clone). As far as we know, nobody has ever performed a successful RF fingerprint duplication [87, 90].

The RF-fingerprinting process usually includes four stages in which signal processing and different algorithms are used. The full RF-fingerprinting process is shown in Fig. 6.2. The signal is first received by the fingerprinting device and converted to digital format using an Analogue-to-Digital Converter (ADC). A transient is then located and its features are extracted. A set of features form an RF fingerprint which can be used for device identification or creating a whole new RF fingerprint for a legitimate device and saving it to the database [90, 91].

Since the use of Bluetooth communication systems and their interconnections via networks have grown rapidly in recent years, secure Bluetooth device identification and pairing are extremely important. As far as we know, there does not exist any complete RF-fingerprinting system for Bluetooth technology, i.e., various researchers have presented different signal processing methods without proposing a complete RF-fingerprinting system for Bluetooth. Thus, we have proposed an *RF-Fingerprint-Based Security Solution for SSP* [90].

RF fingerprinting for Wireless Local Area Networks (WLANs) has been quite well studied [85, 87, 92, 93], but there exist only a few studies [89, 94] concentrating on Bluetooth signals and the identification process. In 2004, Herfurt and Mulliner proposed a method called *BluePrinting* [51] for determining the manufacturer, the device model, and the firmware version of a Bluetooth device (see Sect. 4.1). With this method, the characteristics of the device can be identified, but the method cannot be used for secure Bluetooth device identification [90].

When designing a new wireless security solution, it is very important to be able to answer the following questions [90]:

- What is the main goal of the security solution?
- How can the goal be achieved?
- What other aspects have an effect on the design phase?

In the case of Bluetooth networks, the security solution must be as easy to use as possible: Bluetooth users expect easy-to-use devices, because Bluetooth versions

2.1+EDR, 3.0+HS, and 4.0 support SSP. Moreover, the security solution should be as cheap and reliable as possible. These attributes will also appeal to the Bluetooth device manufacturers [90].

The proposed system is local, i.e., it is used by a single company inside a single building, and the system includes a Bluetooth-enabled server/sensor in a fixed location. The server serves legitimate Bluetooth devices and the only thing required by the legitimate devices is a small program, which takes care of the communication between the legitimate devices and the server. In this way, additional hardware is not required for legitimate Bluetooth devices. Moreover, we feel that this is a very economical solution, since it should have very little effect on the size, power consumption, or price of a Bluetooth device: we will verify this in our future research work with extensive tests [90].

It is entirely up to the user whether she wants to use the system or not: the user can switch the system on/off from the settings of her Bluetooth device. The main goal of device identification can be regarded in two ways. We strongly recommend that the main goal of the system should be the identification of the legitimate devices, not the attacking ones, since the legitimate devices have to be able to communicate with each other in the Bluetooth network. Another option would be detecting the attacking devices, but we feel that this approach is not very efficient in practice [90].

The server must have all legitimate *RF fingerprint–BD_ADDR* pairs (RF fingerprint profiles) stored in its database. The communication with the server will take place in the following way (see Fig. 6.3) [90]:

1. Device B wants to connect with the legitimate device A.
2. The devices use SSP for pairing.
3. Before the actual data exchange, the device A sends a query along with the BD_ADDR of the device B (BD_ADDR$_B$) to the server, which verifies whether or not the BD_ADDR$_B$ is on the list of legitimate devices. Furthermore, the server verifies whether or not the transient captured by the sensor matches the previously stored RF fingerprint profile.
4. Based on the results of step 3, the server sends a connection acceptance or denial notification to the device A. In the case of denial notification, the program installed in the device A performs automatic disconnection with the attacking device. In the case of acceptance notification, the connection is accepted and the Bluetooth devices can start a normal data exchange.

All computationally intensive tasks are performed by the server. From the point of view of users, the acceptance/denial notification must be received within a few seconds, because otherwise users are likely to opt for less secure but more usable options. Thus, the server must be efficient enough to be able to perform the whole RF-fingerprinting process within a few seconds [90].

The server should monitor the Bluetooth network non-stop and also detect hidden Bluetooth devices by using the techniques described in [2, 30, 31] (see Sect. 4.1). In this way, the network administrator will get valuable real-time information about the Bluetooth network: for example, the number of unknown devices in range and the number of possible attack attempts against the network [90].

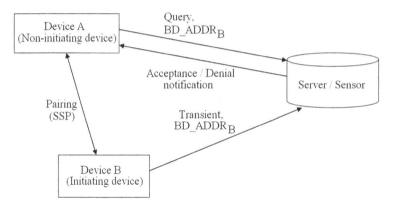

Fig. 6.3 Proposed Bluetooth identification system [90]

The server can be, for example, a Linux-based computer equipped with a Bluetooth protocol analyzer and signal processing capabilities, which are required for transient processing and RF fingerprint formatting. The transient processing will be done in digital form. The server's own security should be extremely strong: the server should be configured only through a wired connection by the Bluetooth network administrator [90].

As described in Sect. 3.1, there are three Bluetooth device classes: class 1, class 2, and class 3. The desired communication range of the system will define both the required transmission power and the sensitivity level of the server's Bluetooth module, which should be substantially better than the minimum requirements of the Bluetooth specification in order to cover large buildings with a single server. This may even require the use of a power amplifier with a class 1 Bluetooth transmitter in addition to an extremely sensitive receiver. It is also worth noting that the Bluetooth network administrator should update the RF fingerprint profiles regularly (e.g., once in a year), because the signal features change as the device ages [90].

6.3 Countermeasures for Bluetooth Devices up to 2.0 + EDR

The following countermeasures can be used to protect Bluetooth versions up to 2.0 + EDR:

- *Increasing user understanding of security issues:* A user should be aware of the Bluetooth security architecture and know how to set it up successfully (see Chap. 2) [2, 53].
- *Updating latest firmware/software on vulnerable Bluetooth devices:* For example, most of the Bluetooth mobile phone firmware/software versions since summer 2004 are assumed to be safe against a BlueBugging attack and several other Bluetooth security attacks, because most mobile phone manufacturers fixed the security

flaws (i.e., flaws in the authentication and data transfer mechanisms) of their bad Bluetooth firmware/software implementations during summer 2004 [2, 53, 61].

- *Data/voice encryption:* All sensitive material should be encrypted with a 128-bit encryption key to prevent its unauthorized use (see Table 3.3 in Sect. 3.2) [2, 53].
- *Private or silent security level:* BD_ADDR should not be public, because it is more difficult for an attacker to synchronize with the piconet's hop sequence when the security level is set as private or silent. However, techniques for finding hidden Bluetooth devices in an average of one minute have been proposed [30] and even implemented [31] (see Sect. 4.1). Thus, this countermeasure only slows down attackers a bit, but does not eliminate any attacks [2, 53].
- *Switching Bluetooth off completely if there is no need to use it for a long time:* For example, a printer's Bluetooth functionality can be switched off when there is no need to use the Bluetooth-enabled printer for a long time [2, 53, 55].
- *Automatic power off capability:* Bluetooth devices with fixed PIN codes should automatically (when possible) turn their power off if no successful connection attempt is made within some predetermined time [2].
- *Minimization of transmit powers:* Sensitive data should be sent using the smallest transmit power possible. On the other hand, this increases the range of susceptibility to jamming (see Table 4.6 in Sect. 4.3) [2, 53].
- *Monitoring sudden increase in transmit powers* (see Sect. 6.1) [2, 83].
- *Careful selection of place:* This is very important especially when two devices meet for the first time and generate initialization keys (see Fig. 2.1 in Chap. 2) [2, 53].
- *Using only long PIN codes:* Sixteen 8-bit character PIN codes should be used when possible (see Sect. 3.3). If a Bluetooth device, such as a headset, a keyboard, or a printer adapter, has a fixed PIN code, it should be as long as possible and as hard as possible to guess [2, 53].
- *Using additional security at application level:* Application layer key exchange and encryption methods can be used as extra security in addition to the Bluetooth built-in security: indeed, our open Public Key Infrastructure (PKI)-based *Mobile Payment System* [95] can be seen as an example of a real-world Bluetooth-enabled system that uses application layer key exchange and encryption methods to secure communication on top of the existing Bluetooth security measures. Although all data exchanged via Bluetooth are encrypted using built-in encryption with 128-bit keys (see Table 3.3 in Sect. 3.2), we use Bluetooth as an untrusted transport medium. All sensitive data are encrypted on the application level. The integrity and freshness of messages is ensured by digital signatures, timestamps, and nonces. Compared with other mobile payment systems, the main advantage of our system is that it does not require any mediator. This reduces the total cost of a payment. Our system utilizes a governmental public key infrastructure, namely Finnish Electronic Identification (FINEID), making it an affordable solution, since administration of the system is provided by the government. Furthermore, as citizens have adopted this system for secure electronic transactions, it has a high level of trustworthiness. Our system is built using Java to gain the best possible portability across device platforms. Our solution provides strong authentication

of communicating parties, integrity of data, non-repudiation of transactions, and confidentiality of communication. Based on the governmental PKI, the system is open to all merchants, financial institutions, and mobile users. More details about our open PKI-based Mobile Payment System can be found in [95]. Another example could be the case when a user has a Bluetooth Access Point with a Print Server providing wireless printing services via Bluetooth: it may be possible to use application layer key exchange and encryption methods as extra security in addition to the Bluetooth built-in security [2, 55, 95].

- *Always requiring a button press from the user prior to accepting the connection establishment:* A Bluetooth connection is accepted only if the user physically confirms the connection establishment by pressing a button [2].

- *Always changing the PIN code without sending a new PIN code via Bluetooth link:* PIN code changing should be done directly using, for example, the Bluetooth keyboard itself without any help from the target computer. This can be done, for example, by pressing a particular button for a predetermined time (for example, 10 s), typing the desired new PIN code, and then pressing the same button again to accept the PIN code storage. It is worth noting that the same new PIN code must also be typed at the computer side using a traditional wired keyboard [2].

- *Switching off all unnecessary SCO/eSCO links:* This countermeasure effectively eliminates SCO/eSCO attacks (see Sect. 4.3) [2, 55].

- *Always requiring an additional Bluetooth-independent reauthentication prior to accessing sensitive information or service:* For example, by requiring an additional Bluetooth-independent reauthentication prior to accepting the establishment of a SCO/eSCO link, SCO/eSCO attacks (see Sect. 4.3) can be effectively eliminated. Another example could be the case when a user has a Bluetooth Access Point with a Print Server: it may be possible to configure the Bluetooth Access Point in such a way that it always requires Bluetooth-independent reauthentication prior to accepting the connection establishment to the Bluetooth printer [2, 55, 61].

- *Printing sensitive information via a traditional cable-based connection to the printer:* The user should not print any sensitive information via Bluetooth. Instead, a traditional cable-based connection to the printer should be used [2, 55].

- *Using an intrusion detection and prevention system:* Various Bluetooth security attacks in progress can be prevented and stopped by monitoring Bluetooth communications to discover such attacks (see Sect. 6.1) [2, 83].

- *Using RF signatures* (see Sect. 6.2) [2, 84–90].

- *Using a portable Bluetooth Direction-Finding device:* It is possible to use a portable Bluetooth Direction-Finding device to determine the area from which an attacking device performs repeated successful connection establishments or other harmful actions. Then the attacking device can be physically found and switched off or even destroyed [2].

- *Keeping a list of suspicious devices in a database* [2, 53].

6.4 Countermeasures for SSP-Enabled Bluetooth Devices

Provided that suitable countermeasures from Sect. 6.3 are in place, the following SSP-specific countermeasures can be used to protect SSP-enabled Bluetooth devices:

- *Force SSP-enabled Bluetooth devices to accept only authenticated link keys:* The simplest and cheapest countermeasure against SSP MITM attacks is to force devices to accept only authenticated link keys. Thus, *we propose that devices should require MITM protection during SSP and enforce it by not accepting unauthenticated link keys*, which are generated only by the Just Works association model. In practice, this can be accomplished as follows. The Bluetooth specification [1] discusses the concept of a "security database" that contains an entry for each service along with the security requirements of that service. Bluetooth protocol stacks commonly include this "security database" function. One of these security requirements should be that the device requires an authenticated link key. If this security requirement is not met, access to the service is not granted. Therefore, this simple countermeasure is up to the discretion of the Bluetooth protocol stack provider, i.e., the countermeasure can be used only if the Bluetooth protocol stack provides a "security database" function.
- *An additional window at the user interface level:* SSP MITM attacks can be prevented at the user interface level. We recommend that an additional window, "The second device has no display and keyboard! Is this true?", should be displayed at the user interface level of SSP when the Just Works association model is to be used. The user is asked to choose either "Proceed" or "STOP". In practice, future Bluetooth specifications should recommend Bluetooth device/software manufacturers to implement this new window as a security improvement of SSP. The advantage of this approach is that the Just Works association model can still be a part of the future Bluetooth SSP specifications without any changes. On the other hand, this countermeasure is a clear trade-off between security and usability [2, 20-23].
- *Just Works as an optional (not mandatory) association model:* Devices that cannot enforce the use of authenticated link keys, use the new window at the user interface level, or NFC as an OOB channel (a better way) should implement their security either in the same way as old Bluetooth devices (versions up to 2.0+EDR) do, or not use Bluetooth security at all (if no sensitive data is exchanged). In this way, the implementation of the Just Works association model can be made optional and perhaps even removed altogether from the Bluetooth SSP specification. The one advantage of this approach is that it eliminates all MITM attacks against the Just Works association model. Moreover, if the Just Works association model is not supported in future Bluetooth devices, it will not be possible to force victim devices to use it [2, 21-23].
- *OOB as a mandatory association model:* Future Bluetooth specifications should make OOB a mandatory association model in order to improve the security and usability of SSP. However, it is likely that such a radical change in the specification will not be possible at once. Therefore, future Bluetooth specifications should at

least recommend the use of an OOB channel (for example, NFC) to all Bluetooth device manufacturers [2, 21-23].

- *Using SSP's OOB channel only in a secure environment:* SSP's OOB channel should only be used in a location where an attacker cannot have any kind of visual contact to the victim devices [2, 22, 23].

Chapter 7
New Practical Attack

Based on our findings and practical experiments on Big NAK attacks as well as our research work on proposing an Intrusion Detection and Prevention System for Bluetooth networks (see Sect. 6.1), we propose a new attack called a *Big POLL attack*, which works against all existing Bluetooth versions, i.e., Bluetooth versions 1.0A–4.0.

In the attack, the attacker keeps victim devices (piconet slaves) busy all the time by sending repeated POLL packets to them so that they will not go into sleep or low-power mode. The Big POLL attack is possible because during a normal piconet operation the master device can use POLL packets to check that slave devices are still alive (i.e., up and running), and slave devices must always respond to a POLL packet sent by the master device.

The attacker needs to discover the BD_ADDRs of all piconet devices (i.e., the BD_ADDR of the master device and BD_ADDRs of slave devices), impersonate the piconet master (i.e., duplicate its BD_ADDR), and start sending POLL packets to all the piconet slaves.

Let us assume the following attack scenario:

1. The attacker uses a Brute-Force BD_ADDR Scanning attack, a Bluetooth protocol analyzer, techniques for finding hidden Bluetooth devices in an average of one minute, or 79 Bluetooth receivers in parallel (see Sect. 4.1) to discover the hidden (non-discoverable) BD_ADDRs of the piconet master and all piconet slaves.
2. The attacker impersonates the piconet master by duplicating its BD_ADDR. Some commercially available Bluetooth protocol analyzers, such as LeCroy BTTracer/Trainer [42], support the BD_ADDR duplication feature. Therefore, this is not a problem for the attacker.
3. The attacker starts sending POLL packets non-stop to all piconet slaves, thus keeping them busy all the time.

In our attack, the attacker consumes the batteries of the victim devices (piconet slaves) rather quickly. Moreover, the piconet slaves do not receive the legitimate piconet services within a reasonable time, because the attacking device keeps them

K. Haataja et al., *Bluetooth Security Attacks*, SpringerBriefs in Computer Science, DOI: 10.1007/978-3-642-40646-1_7, © The Author(s) 2013

busy all the time. It is worth noting that the attacker does not need to witness the initial pairing process between the victim devices or intercept any random numbers sent via air: the only information required is the BD_ADDRs of the victim devices.

Big POLL attacks are not normally very dangerous, because the attacker does not steal any information from the target devices. However, these kinds of attacks can be very annoying if the attacker uses them non-stop to deny the legitimate piconet devices access to the piconet services, or at least in such a way that they have considerably slowed throughput. Moreover, the attacker can use these kinds of attacks to mislead the target devices in such a way that they delete previously stored link keys so that the initial pairing process is restarted.

The only feasible countermeasure against a Big POLL attack seems to be the following: *a Bluetooth network should use our Intrusion Detection and Prevention System* (see Sect. 6.1) or some other intrusion detection and prevention system that is capable of detecting a Big POLL attack. Moreover, BD_ADDRs can be set as hidden (non-discoverable), but as we discussed earlier, this only slows down the attacker a bit and thus it is not a very convincing countermeasure.

Chapter 8
Conclusion and Future Work

In this book, reasons for Bluetooth network vulnerabilities were explained and a literature-review-based comparative analysis of Bluetooth security attacks over the past ten years (2001–2011), including our own Bluetooth security attacks, was provided. In addition, countermeasures against these attacks were explained based on a literature review and a new practical countermeasure for Bluetooth SSP was proposed. Moreover, a novel attack that works against all existing Bluetooth versions was proposed.

Overall security in Bluetooth networks is based on the security of the Bluetooth medium, the security of Bluetooth protocols, and the security parameters used in Bluetooth communication. There are several weaknesses in the Bluetooth medium, Bluetooth protocols, and Bluetooth security parameters which can significantly weaken the overall security of Bluetooth networks. Currently, weaknesses in the Bluetooth security parameters seem to be the biggest problem in Bluetooth security.

The current level of security is insufficient in many Bluetooth devices on the market, as this book clearly shows. Many kinds of Bluetooth devices have very short, often only four-digit, fixed PIN codes. This is clearly a big security risk. Therefore, Bluetooth device manufacturers should take security issues more seriously. In addition, users' understanding of security issues is very important for protecting sensitive data against eavesdroppers and hackers. Moreover, many users have no idea how to configure their Bluetooth devices' security settings correctly. Application layer key exchange and encryption methods can also be used as extra security in addition to the Bluetooth built-in security.

Many attacks, such as Reflection attacks and Interception of Packets attacks, are possible because encryption is not used by default in many kinds of Bluetooth devices. We strongly recommend that Bluetooth factory settings should enable encryption by default, because many users do not know about its existence or do not know how to set it up successfully. Another possibility is to set Bluetooth encryption as mandatory. This of course would require minor changes to the Bluetooth specification, and it would mean that older Bluetooth devices would also have to use encryption

K. Haataja et al., *Bluetooth Security Attacks*, SpringerBriefs in Computer Science, DOI: 10.1007/978-3-642-40646-1_8, © The Author(s) 2013

or communication with new devices would not be possible. On the other hand, if backward-compatibility with old Bluetooth devices must be guaranteed, mandatory encryption is not possible.

Many attacks, such as Off-Line PIN Recovery attacks and On-Line PIN Cracking attacks, are also possible because many kinds of Bluetooth devices have very short fixed PIN codes containing only digits. We strongly recommend that the allowed sixteen 8-bit character PIN codes should always be used when possible.

Dozens of attacks are also possible because many kinds of Bluetooth devices have public security level as a fixed factory setting, so they are always discoverable. We strongly recommend that the security level of Bluetooth devices should not be public by default or as the fixed factory setting. The user should at least have the option of changing the default factory security level setting somehow. However, in the near future, when techniques for finding hidden Bluetooth devices in an average of one minute are actually ported into the firmware of a standard $30 Bluetooth USB dongle, the private security level will no longer provide any significant protection.

Since there are billions of Bluetooth devices in use without SSP's improved security features, malicious security violations are not expected to decrease in the near future. On the contrary, these old Bluetooth devices will be sold for many years to come, thus making security concerns even more alarming.

SSP has gone through a series of reviews by experts and the released version generally does good work in improving the security of Bluetooth pairing. However, MITM attacks against SSP are still possible, as this book clearly shows. Therefore, the Bluetooth security architecture needs to be further updated to prevent these new threats.

It is difficult to create a protocol which caters to all possible types of wireless devices, as the security of the protocol is likely to be limited by the capabilities of the least powerful or the least secure device type. Most Bluetooth security attacks are based on exactly this problem.

In general, MITM attacks are hard to prevent in wireless networks. By far the best way to stop the attacks is to use SSP's OOB channel in such a way that an attacker cannot have visual contact to the victim devices. Moreover, the usability of the OOB channel is of great importance: if wires must be used for pairing wireless devices, the user is likely to opt for less secure but more usable options.

Since we have now covered all noteworthy Bluetooth security attacks, including our own attacks, during the past 10 years (2001–2011) in this book, it is also valuable to summarize the dangerousness and practical relevance of the attacks: thus, Fig. 8.1 summarizes all attacks described in this book by providing dangerousness and practical relevance figures.

The problems we want to investigate in our future research work are concerned with the following issues:

1. Bluetooth is a relatively new wireless technology and therefore new attacks against Bluetooth security are likely to be found. We want to further investigate Bluetooth security weaknesses and propose countermeasures against new attacks.

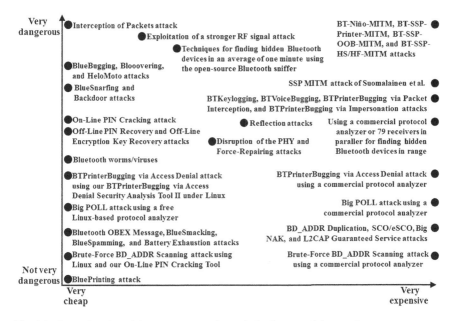

Fig. 8.1 Our estimation of dangerousness and practical relevance of the attacks

2. Issues related to Bluetooth user experience (ease of use) have become more and more important in recent years. Therefore, we want to investigate how an enhanced user experience will affect Bluetooth security in various scenarios, including social aspects and user acceptance/habits in security management. Moreover, we want to devise best practices depending on the risk analysis within each scenario.

3. Since we have already proposed a new efficient Intrusion Detection and Prevention System for Bluetooth networks, we want to implement a working prototype of such a system and also analyze its efficiency.

4. We want to extend our Bluetooth intrusion detection/prevention-related research to cover also RF-fingerprinting techiques, because we feel that the use of RF fingerprints could be the future of secure Bluetooth communications. Thus, we want to implement a working prototype of our proposed system (see Sect. 6.2) and also analyze its efficiency. After that, we plan to extend the current local approach into a global approach in which the RF fingerprint profiles will be stored to a server via an Internet connection and thus they can be used in all locations. In this way, a multinational company can use the same RF fingerprint profiles all around the world.

5. Because cheap Bluetooth devices with an adjustable BD_ADDR are readily available, tools for modifying official firmware have been released, techniques for finding hidden Bluetooth devices in an average of one minute have been invented, and an open-source Bluetooth sniffer for Linux environments has been released,

we want to continue our practical Bluetooth security research under Linux using these new tools. We also want to further develop the existing open-source Bluetooth sniffer to include the BD_ADDR duplication feature and the graphical user interface in order to make it user-friendly.

6. Since nowadays it is possible to acquire the hardware required for MITM attacks, we want to make practical implementations of all existing Bluetooth MITM attacks. Moreover, we want to analyze the results of the practical experiments, draw conclusions, and propose practical countermeasures based on our findings.

7. Since there are many new emerging wireless technologies, such as ZigBee and Ultra-Wideband (UWB), which are quite similar to Bluetooth technology, it is expected that our Bluetooth-security-related research work can be quite easily extended to cover the security of these new technologies. Therefore, we want to investigate how various Bluetooth security attacks and their countermeasures can be ported to support ZigBee and UWB technologies. In fact, we are currently conducting practical research work on ZigBee security attacks.

8. We feel that the use of steganography [96–101] could be one potential solution for securing Bluetooth communications in the future: in fact, we have performed some research work on steganography [96, 97] and we are also currently conducting research work on the computational aspects of watermarking and steganography to allow secure authentication between communicating Bluetooth devices, i.e., we plan to embed certain messages into digital images by using digital watermarking and/or steganography to hide the existence of the messages, allowing secure extraction of these embedded messages only by the legitimate recipient.

Bluetooth security intimately depends on general problems of ad hoc network security, on physical aspects of protecting wireless networks, on cryptographical solutions to key distribution without Trusted Third Party or Certification Authority infrastructure, and on application layers. Research in this area combines various skills and techniques, requires cooperation with other researchers, and also requires a certain infrastructure.

References

1. Bluetooth SIG, Bluetooth specifications 1.0A-4.0 (2013), https://www.bluetooth.org/Technical/Specifications/adopted.htm. Accessed 2 May 2013
2. K. Haataja, Security threats and countermeasures in Bluetooth-enabled systems, Ph.D. Dissertation, Department of Computer Science, University of Kuopio, 6 Feb 2009
3. Bluetooth SIG, Bluetooth technology bringing unprecedented functionality to wireless devices in fitness, health, home, and automotive (2013), http://www.bluetooth.com/Pages/Press-Releases-Detail.aspx?ItemID=126. Accessed 2 May 2013
4. PRWeb, Bluetooth-enabled device shipments expected to exceed 2 billion in 2013, says In-Stat (2013), http://www.prweb.com/releases/In-Stat/Bluetooth-Enabled-Device/prweb8723683.htm. Accessed 2 May 2013
5. IEEE Registration Authority, IEEE public OUI and company_id assignments (2013), http://standards.ieee.org/regauth/oui/oui.txt. Accessed 2 May 2013
6. J. Massey, G. Khachatrian, M. Kuregian, SAFER+, in *Proceedings of the 1st NIST Advanced Encryption Standard Candidate Conference*, Ventura, California, USA, 20–22 August 1998
7. NIST, Advanced Encryption Standard (AES): the Federal Information Processing Standards publication 197 (2013), http://csrc.nist.gov/publications/fips/fips197/fips-197.pdf. Accessed 2 May 2013
8. Y. Shaked, A. Wool, Cracking the Bluetooth PIN, in *Proceedings of the 3rd ACM International Conference on Mobile Systems, Applications, and Services*, Seattle, Washington, USA, 6–8 June 2005, pp. 39–50
9. J. Zhao, M. Wang, J. Chen, Y. Zheng, New impossible differential attack on SAFER+ and SAFER++, in *Proceedings of the 5th International Conference on Information Security and Cryptology (ICIS2012)*, Lecture Notes in Computer Science, vol. 7839 (Springer, 2013), pp. 170–183
10. C. Pfleeger, *Security in Computing*, 3rd edn. (Prentice Hall, Upper Saddle River, 2003)
11. W. Stallings, *Cryptography and Network Security: Principles and Practice*, 3rd edn. (Prentice Hall, Upper Saddle River, 2003)
12. R. Morrow, *Bluetooth: Operation and Use* (McGraw-Hill, New York, 2002)
13. M. Jakobsson, S. Wetzel, Security weaknesses in Bluetooth, in *CT-RSA2001: Topics in Cryptology*, Lecture Notes in Computer Science, vol. 2020, (Springer, 2001), pp. 176–191
14. D. Kügler, Man-in-the-middle attacks on Bluetooth, in *Proceedings of the 7th International Conference on Financial Cryptography*, Lecture Notes in Computer Science, vol. 2742 (Springer, 2003), pp. 149–161
15. A. Levi, E. Cetintas, M. Aydos, C. Koc, M. Caglayan, Relay attacks on Bluetooth authentication and solutions, in *Proceedings of the 19th International Conference on Computer and*

Information Sciences (ISCIS2004), Lecture Notes in Computer Science, vol. 3280 (Springer, 2004), pp. 278–288

16. B. Schneier, *Applied Cryptography: Protocols, Algorithms, and Source Code in C*, 2nd edn. (Wiley, New York, 1996)
17. L. Kohnfelder, Towards a practical public-key cryptosystem, Bachelor's Thesis, MIT, May 1978
18. J. Buchmann, *Introduction to Cryptography* (Springer, New York, 2001)
19. J. Suomalainen, J. Valkonen, N. Asokan, Security associations in personal networks: a comparative analysis, in *Proceedings of the 4th European Workshop on Security and Privacy in Ad-hoc and Sensor Networks*, Lecture Notes in Computer Science, vol. 4572 (Springer, 2007), pp. 43–57
20. K. Hyppönen, K. Haataja, "Niño", man-in-the-middle attack on Bluetooth secure simple pairing, in *Proceedings of the 3rd IEEE International Conference in Central Asia on Internet, the Next Generation of Mobile, Wireless, and Optical Communications Networks*, Tashkent, Uzbekistan, 26–28 Sept 2007
21. K. Haataja, K. Hyppönen, Man-in-the-middle attacks on Bluetooth: a comparative analysis, a novel attack, and countermeasures, in *Proceedings of the 3rd IEEE International Symposium on Communications, Control, and Signal Processing*, St. Julians, Malta, 12–14 March 2008
22. K. Haataja, P. Toivanen, Practical man-in-the-middle attacks against Bluetooth secure simple pairing, in *Proceedings of the 4th IEEE International Conference on Wireless Communications, Networking, and Mobile Computing*, Dalian, China, 12–14 Oct 2008
23. K. Haataja, P. Toivanen, Two practical man-in-the-middle attacks on Bluetooth secure simple pairing and countermeasures. IEEE Trans. Wireless Commun. 9(1), 384–392 (2010)
24. J. Barnickel, J. Wang, U. Mayer, Implementing an attack on Bluetooth 2.1+ secure simple pairing in passkey entry mode, in *Proceedings of the 11th IEEE International Conference on Trust, Security, and Privacy in Computing and Communications*, Liverpool, UK, 25–27 June 2012, pp. 17–24
25. R. Anderson, *Security Engineering: A Guide to Building Dependable Distributed Systems* (Wiley, New York, 2001)
26. H. Cheung, How to: building a BlueSniper rifle, part 1 (2013), http://www.smallnetbuilder. com/content/view/24256/98. Accessed 2 May 2013
27. H. Cheung, How to: building a BlueSniper rifle, part 2 (2013), http://www.smallnetbuilder. com/content/view/24228/98. Accessed 2 May 2013
28. M. Moser, Busting the Bluetooth myth: getting RAW access (2013), http://packetstorm. wowhacker.com/papers/wireless/busting_bluetooth_myth.pdf. Accessed 2 May 2013
29. Darkircop, CSR sniffer: firmware assembler and disassembler (2013), http://darkircop.org/ bt/bt.tgz. Accessed 2 May 2013
30. D. Spill, A. Bittau, BlueSniff: Eve meets Alice and Bluetooth, in *Proceedings of the 1st USENIX Workshop on Offensive Technologies*, Boston, 6 August 2007
31. D. Spill, A. Bittau, BlueSniff source codes (2013), http://www.cs.ucl.ac.uk/staff/a.bittau/gr-bluetooth.tar.gz. Accessed 2 May 2013
32. Project Ubertooth, An open source 2.4 GHz wireless development platform suitable for Bluetooth experimentation (2013), http://ubertooth.sourceforge.net. Accessed 2 May 2013
33. M. Herfurt, Detecting and attacking Bluetooth-enabled cellphones at the Hannover fairground (2013), http://trifinite.org/Downloads/BlueSnarf_CeBIT2004.pdf. Accessed 2 May 2013
34. A. Laurie, B. Laurie, Serious flaws in Bluetooth security lead to disclosure of personal data (2013), http://www.oocities.org/h4k3r5. Accessed 2 May 2013
35. O. Whitehouse, @Stake: where security & business intersect (2013), http://cansecwest.com/ csw04archive.html. Accessed 2 May 2013
36. Bluediving Project, Bluediving: next generation Bluetooth security tool (2013), http:// bluediving.sourceforge.net. Accessed 2 May 2013
37. BlueZ Project, BlueZ: official linux Bluetooth protocol stack (2013), http://www.bluez.org. Accessed 2 May 2013

38. Infrared Data Association, IrDA object exchange protocol (OBEX) specifications (2013), http://www.irda.org. Accessed 2 May 2013
39. C. Gehrmann, J. Persson, B. Smeets, *Bluetooth Security* (Artech House, Boston, 2004)
40. O. Whitehouse, RedFang: Bluetooth discovery tool (2013), http://www.securiteam.com/tools/5JP0I1FAAE.html. Accessed 2 May 2013
41. K. Haataja, Two practical attacks against Bluetooth security using new enhanced implementations of security analysis tools, in *Proceedings of the IASTED International Conference on Communication, Network, and Information Security*, Phoenix, Arizona, USA, 14–16 Nov 2005, pp. 13–18
42. LeCroy, LeCroy HCI tracer (2013), http://www.lecroy.com/protocolanalyzer/protocoloverview.aspx?seriesid=103&capid=103&mid=511. Accessed 2 May 2013
43. Nokia, User's guide: Nokia 6310i (2013), http://nds1.nokia.com/phones/files/guides/6310i_usersguide_en.pdf. Accessed 2 May 2013
44. LeCroy, CATC scripting language reference manual for CATC Bluetooth analyzers: manual version 1.21 (2013), http://cdn.lecroy.com/files/manuals/btcsl_d121.pdf. Accessed 2 May 2013
45. A. Laurie, M. Holtmann, M. Herfurt, Hacking Bluetooth-enabled mobile phones and beyond: full disclosure, in *Proceedings of the 21st Chaos Communication Congress* (Berliner Congress Center, Berlin, 2004), pp. 27–29
46. K. Sapronov, Bluetooth, Bluetooth security, and New Year war-nibbling (2013), http://www.viruslist.com/en/analysis?pubid=181198286. Accessed 2 May 2013
47. K. Sapronov, War-nibbling (2007), http://www.viruslist.com/en/analysis?pubid=204791928. Accessed 2 May 2013
48. K. Spencer, Taking a peek inside your mobile (2013), http://news.bbc.co.uk/2/hi/technology/3642627.stm. Accessed 2 May 2013
49. Ettus Research, Universal software radio peripheral (2013), http://www.ettus.com. Accessed 2 May 2013
50. GNU Radio Project, GNU radio: the GNU software radio (2013), http://gnuradio.org/redmine/projects/gnuradio/wiki. Accessed 2 May 2013
51. M. Herfurt, C. Mulliner, BluePrinting: remote device identification based on Bluetooth finger-printing techniques (2013), http://trifinite.org/Downloads/Blueprinting.pdf. Accessed 2 May 2013
52. C. Mulliner, M. Herfurt, BluePrinting (2013), http://trifinite.org/trifinite_stuff_blueprinting.html. Accessed 2 May 2013
53. K. Haataja, Bluetooth security threats and possible countermeasures, in *Proceedings of the Annual Finnish Data Processing Week at the University of Petrozavodsk on Advances in Methods of Modern Information Technology*, vol. 6, Petrozavodsk, Russia (2005), pp. 116–150
54. Viha, Internet relay chat (2013), http://www.irc.org. Accessed 2 May 2013
55. K. Haataja, Three practical Bluetooth security attacks using new efficient implementations of security analysis tools, in *Proceedings of the IASTED International Conference on Communication, Network, and Information Security*, Berkeley, California, USA, 24–26 Sept 2007, pp. 101–108
56. Nokia, User's guide for the wireless headset (HDW-2) (2013), http://nds1.nokia.com/phones/files/guides/Wireless_Headset_hdw2_en.pdf. Accessed 2 May 2013
57. Nokia, Nokia wireless headset (HS-26W) user guide (2013), http://nds2.nokia.com/files/support/apac/phones/guides/HS26W_APAC_UG_en.pdf. Accessed 2 May 2013
58. Sony Ericsson, Bluetooth headset HBH-610 user guide (2013), http://i.smartphone.ua/docs/instr_btg/instr_sony-ericsson-hbh-610a_eng.pdf. Accessed 2 May 2013
59. Epox, Epox communication BT-DG07A+ description (2013), http://www.comsysir.com/Product/detail.aspx?ProductID=160&LangStatu. Accessed 2 May 2013
60. SecureList, Nokia 6310i OBEX message denial-of-service (2013), http://www.securelist.com/en/advisories/10827. Accessed 2 May 2013

61. K. Haataja, Bluetooth network vulnerability to disclosure, integrity, and denial-of-service attacks, in *Proceedings of the Annual Finnish Data Processing Week at the University of Petrozavodsk on Advances in Methods of Modern Information Technology*, vol. 7, Petrozavodsk, Russia (2006), pp. 63–103

62. A. Laurie, M. Holtmann, M. Herfurt, BlueSmack (2013), http://trifinite.org/trifinite_stuff_bluesmack.html. Accessed 2 May 2013

63. C. Mulliner, BlueSpam (2013), http://www.mulliner.org/palm/bluespam.php. Accessed 2 May 2013

64. A. Laurie, BlueBug (2013), http://trifinite.org/trifinite_stuff_bluebug.html. Accessed 2 May 2013

65. A. Oberritter, btxml: mobile phone backup tool using Bluetooth (2013), http://www.saftware.de/bluetooth/btxml.c. Accessed 2 May 2013

66. M. Herfurt, Blooover (2013), http://trifinite.org/trifinite_stuff_blooover.html. Accessed 2 May 2013

67. M. Herfurt, Blooover II (2013), http://trifinite.org/trifinite_stuff_bloooverii.html. Accessed 2 May 2013

68. A. Laurie, HeloMoto (2013), http://trifinite.org/trifinite_stuff_helomoto.html. Accessed 2 May 2013

69. K. Haataja, New practical attack against Bluetooth security using efficient implementations of security analysis tools, in *Proceedings of the IASTED International Conference on Communication, Network, and Information Security*, Berkeley, California, USA, 24–26 Sept 2007, pp. 134–142

70. Conceptronic, Conceptronic Bluetooth printer adapter (2013), http://www.conceptronic.net. Accessed 2 May 2013

71. Bona Computech, Mentor Bluetooth printer adapter (2013), http://www.acesuppliers.com/Supplier_Company/BONA-COMPUTECH-Co-Ltd_Directory_3255751220051419245156416.html. Accessed 2 May 2013

72. Tecom, Tecom Bluetooth printer adapter (2013), http://www.deltaco.se/support/infoSE/bt3051.pdf. Accessed 2 May 2013

73. Belkin International, Belkin Bluetooth printer adapter (2013), http://www.amazon.co.uk/Belkin-Bluetooth-USB-Printer-Adapter/dp/tech-data/B0001Q17NW/ref=de_a_smtd/203-7526837-1808712. Accessed 2 May 2013

74. Frontline, FTS4BT wireless Bluetooth protocol analyzer & packet sniffer (2013) http://www.fte.com/products/FTS4BT-01.asp. Accessed 2 May 2013

75. Reuters, Bluetooth viruses (2013), http://www.dailywireless.org/2005/02/04/bluetooth-viruses. Accessed 2 May 2013

76. J. Jackson, S. Creese, Virus propagation in heterogeneous Bluetooth networks with human behaviors. IEEE Trans. Dependable Secure Comput. **9**(6), 930–943 (2012)

77. F-Secure Corporation, F-secure virus descriptions: Cabir (2013), http://www.f-secure.com/v-descs/cabir.shtml. Accessed 2 May 2013

78. F-Secure Corporation, F-secure virus descriptions: Skulls.D (2013), http://www.f-secure.com/v-descs/skulls_d.shtml. Accessed 2 May 2013

79. F-Secure Corporation, F-secure virus descriptions: Lasco.A (2013), http://www.f-secure.com/v-descs/bluetooth-worm_symbos_lasco_a.shtml. Accessed 2 May 2013

80. M. Velasco, Marcos Velasco Security (2013), http://www.velasco.com.br. Accessed 2 May 2013

81. E. Uzun, K. Karvonen, N. Asokan, Usability analysis of secure pairing methods, in *Proceedings of the 11th International Conference on Financial Cryptography and 1st International Confernce on Usable Security, Lowlands, Scarborough, Trinidad/Tobago*, 15–16 Feb 2007

82. N. Borisov, I. Goldberg, D. Wagner, Intercepting mobile communications: the insecurity on 802.11, in *Proceedings of the 7th ACM Annual International Conference on Mobile Computing and Networking*, Rome, Italy, 16–21 July 2001, pp. 180–189

83. K. Haataja, New efficient intrusion detection and prevention system for Bluetooth networks, in *Proceedings of the ACM International Conference on Mobile, Wireless Middleware, Operating Systems, and Applications*, Innsbruck, Austria, 12–15 Feb 2008

84. J. Shandle, University research aims at more secure Wi-Fi (2013), http://www.informationweek.com/university-research-aims-at-more-secure/192501293. Accessed 2 May 2013

85. M. Barbeau, J. Hall, E. Kranakis, Detecting impersonation attacks in future wireless and mobile networks, in *Proceedings of the 1st International Workshop on Secure Mobile Ad-hoc Networks and Sensors*, Lecture Notes in Computer Science, vol. 4074 (Springer, 2006), pp. 80–95

86. J. Franklin, D. McCoy, P. Tabriz, V. Neagoe, J. Randwyk, D. Sicker, Passive data link layer 802.11 wireless device driver fingerprinting, in *Proceedings of the 15th USENIX Security Symposium*, Vancouver, British Columbia, Canada, 31 July–4 August 2006

87. O. Ureten, N. Serinken, Wireless security through RF fingerprinting. Can. J. Electr. Comput. Eng. **32**(1), 27–33 (2007)

88. O. Ureten, N. Serinken, Bayesian detection of radio transmitter turn-on transients, in *Proceedings of the IEEE Nonlinear Signal and Image Processing Conference*, Antalya, Turkey, June 1999, pp. 830–834

89. J. Hall, M. Barbeau, E. Kranakis, Detecting rogue devices in Bluetooth networks using radio frequency fingerprinting, in *Proceedings of the IASTED International Conference on Communications and Computer Networks*, Lima, Peru, 4–6 Oct 2006

90. S. Pasanen, K. Haataja, N. Päivinen, P. Toivanen, New efficient RF fingerprint-based security solution for Bluetooth secure simple pairing, in *Proceedings of the 43rd IEEE Hawaii International Conference on System Sciences*, Koloa, Kauai, Hawaii, 5–8 Jan 2010

91. K. Rasmussen, S. Capkun, Implications of radio fingerprinting on the security of sensor networks, in *Proceedings of the 3rd IEEE International Conference on Security and Privacy in Communications Networks*, Nice, France, 17–20 Sept 2007, pp. 331–340

92. V. Brik, S. Banerjee, M. Gruteser, S. Oh, Wireless device identification with radiometric signatures, in *Proceedings of the 14th ACM International Conference on Mobile Computing and Networking*, San Francisco, California, USA, 14–19 Sept 2008, pp. 116–127

93. D. Loh, C. Cho, C. Tan, R. Lee, Identifying unique devices through wireless fingerprinting, in *Proceedings of the 1st ACM Conference on Wireless Network Security*, Alexandria, Virginia, USA, 31 March–2 April 2008, pp. 46–55

94. J. Hall, M. Barbeau, E. Kranakis, Detection of transient in radio frequency fingerprinting using signal phase, in *Proceedings of the IASTED International Conference on Wireless and Optical Communications*, Banff, Canada, 14–16 July 2003, pp. 13–18

95. M. Hassinen, K. Hyppönen, K. Haataja, An open PKI-based mobile payment system, in *Proceedings of the International Conference on Emerging Trends in Information and Communication Security*, Lecture Notes in Computer Science, vol. 3995 (Springer, 2006), pp. 86–100

96. A. Kaarna, P. Toivanen, Digital watermarking of spectral images in PCA/wavelet-transform domain, in *Proceedings of the IEEE International Geoscience and Remote Sensing Symposium*, vol. 6, Toulouse, France, 21–25 July 2003, pp. 3564–3567

97. A. Kaarna, P. Toivanen, K. Mikkonen, Watermarking spectral images through the PCA transform, in *Proceedings of the IS&T Image Processing, Image Quality, and Image Capture Systems Conference*, vol. 6, Rochester, New York, USA, 13 May 2003, pp. 220–225

98. C. Podilchuk, E. Delp, Digital watermarking: algorithms and applications. IEEE Signal Process. Mag. **8**(4), 33–46 (2001)

99. D. Artz, Digital steganography: hiding data within data. IEEE Internet Comput. **5**(3), 75–80 (2001)

100. K. Satish, T. Jayakar, C. Tobin, K. Madhavi, K. Murali, Chaos based spread spectrum image steganography. IEEE Trans. Consum. Electron. **50**(2), 587–590 (2004)

101. I. Mokowitz, G. Longdon, L. Chang, A new paradigm hidden in steganography, in *Proceedings of the ACM New Security Paradigms Workshop*, Ballycotton, County Cork, Ireland, Sept 2000, pp. 41–50